Children Reading MATHEMATICS

Edited by
Hilary Shuard & Andrew Rothery

Language & Reading in Mathematics Group

Rosemary Hubbard, Sheffield City Polytechnic

Spencer Instone, School Mathematics Project (formerly Finham Park Comprehensive School, Coventry)

Jim Noonan, Education Office, Guildford (formerly St Vincent School, Gosport)

Andrew Rothery, Worcester College of Higher Education

Hilary Shuard, Homerton College, Cambridge

John Murray

First published 1984
by John Murray (Publishers) Ltd
50 Albemarle Street
London W1X 4BD

Reprinted 1988, 1992

Printed in Great Britain by Athenaeum Press, Newcastle upon Tyne

British Library Cataloguing in Publication Data

Children reading mathematics.
1. Mathematics—1961—
I. Shuard, Hilary   II. Rothery, Andrew
510   QA39.2

ISBN 0-7195-4093-3

Preface

This book has arisen from the work of the Language & Reading in Mathematics Group, which was formed in 1978 following meetings organised by the British Society for the Psychology of Learning Mathematics. Its early activities were supported by a grant from the Mathematical Education Section of NATFHE and one of the group's first projects was to prepare a working paper, *Children Reading Maths*, which provided background information in the field of reading mathematics and a report of the group's initial research.

Children Reading Maths was published informally by the group in 1980. The present book, *Children Reading Mathematics*, is a development of this earlier publication and represents a considerable improvement and extension of the material we had then prepared.

The Language & Reading in Mathematics Group has studied the relation between the pupil and the written materials so commonly used in mathematics lessons. This interest has been shared by teachers, advisers, authors and researchers, who have all been stimulated and helped by discussing their problems and ideas with each other.

We have found a number of different professional concerns expressed by those interested in the reading of mathematics materials. Many wish to improve the way books are used, many wish to improve the way they write worksheets and books, some would like to help teach children how to read mathematics materials more skilfully. Teachers frequently would like to develop criteria by which they can evaluate textbooks, and these would include a matching of the readability of the book with the reading ability of the pupil. Children's difficulties in mathematics are often related to difficulties in understanding what is written and what is being asked of them.

We hope *Children Reading Mathematics* will provide the reader with the kind of assistance and inspiration we have received through our group's discussions and activities. We also hope that other groups who meet to work on this topic will be provoked by our book into having useful arguments, and that they will make further progress by taking up (or objecting to) our suggestions and ideas.

Acknowledgements

We would like to thank Michael Holt and Graham Ruddock for their participation as co-authors in the writing of the earlier publication, *Children Reading Maths*, from which this book has grown.

Our thanks also extend to those members of the Language & Reading in Mathematics Group whose participation in the group's activities have stimulated so much of our thinking, and to Sr Timothy Pinner for help in the preparation of some material for the book.

We are also grateful to the large number of colleagues whose comments and advice on our manuscript at its various stages of production have proved extremely valuable; and to Kath Hart and Geoffrey Howson for their comments on the final draft.

The rest of the writing team would like to thank Hilary Shuard for her part in writing this book. She has been a source of enthusiasm, imagination, knowledge and energy which has provided our work with momentum and direction. Hilary has contributed a substantial proportion of the book and given considerable assistance to us in the writing of our own sections. She has done more than her fair share of both the writing and editing of this book.

Contents

1 Reading and mathematics

The importance of reading in mathematics

Written communication is a major component of the methods of teaching mathematics used in most schools today. Textbooks, worksheets and workcards, either commercially produced or home-made, form an integral part of the resources which teachers of mathematics use with their pupils. A printed page of mathematical text may communicate comparatively easily with children, or it may fail to communicate; it is important for teachers of mathematics in both primary and secondary schools to recognise whether children are likely to be able to read the page easily.

In this book, we shall use *reading* in a very wide sense. By *reading* we shall mean the whole process by which a pupil examines the written word and the pictorial material, and obtains its meaning. In mathematics, for us,

reading is 'getting the meaning from the page'.

The process of reading is influenced by the style of writing, the graphic images used and the presentation of the page. The style of mathematical writing is strikingly different from the styles found in non-mathematical text; we shall see that the nature of mathematical text poses special problems for the reader and demands that he acquire special reading skills.

The Bullock report (1975) stated:

> We must convince the teacher of History and Science, for example, that he has to understand the process by which his pupils take possession of the historical or scientific information that is offered to them; and such an understanding involves his paying particular attention to the part language plays in learning.

After the publication of the Bullock report, all teachers were urged to think of themselves as teachers of reading, and mathematics teachers were encouraged to appreciate that language factors affect mathematics performance.

Readability

The concept of readability is encountered through the common experience of teachers, who find that some text passages are easier for pupils to read and understand than others. Dale and Chall (1948) defined readability thus:

> In the broadest sense, readability is the sum total (including interactions) of all those elements within a given piece of printed material that affects the success which a group of readers have with it. The success is the extent to which they understand it, read it at optimum speed and find it interesting.

This definition both alludes to aspects of the text, such as vocabulary, syntax, and layout, and refers to the reader's comprehension, fluency and enjoyment. From about 1948 onwards, both in Britain and elsewhere, a number of *readability formulae* were developed. They were intended to help teachers to assess the readability of particular pieces of text which their pupils might be asked to read. The application of a typical readability formula involves taking sample pages from a text, and counting the average word length and the average sentence length. These measures are then used to estimate the age of child for whom the text is suitable. Many textbooks in all subject areas have been examined using such formulae, and the results which emerged have been used to support the contention that subject specialist teachers neglect the importance of readability in the reading materials used in their subject areas. The situation in mathematics is complicated, however, by the many special features of mathematics text, such as formulae, graphs and diagrams.

Kane (1967) introduced the terms *Ordinary English* and *Mathematical English* to help stress the special nature of written mathematics.

> Mathematical English (ME) is a hybrid language. It is composed of ordinary English (OE) commingled with various brands of highly stylised formal symbol systems. The mix of these two kinds of language varies greatly from elementary school texts to books written for graduate students.

In a later article, Kane (1970) remarked:

> Mathematical english and OE are sufficiently dissimilar that they require different skills and knowledge on the part of readers to achieve appropriate levels of reading comprehension.

In the two articles quoted, Kane pointed out that the differences between OE and ME make it impossible to apply standard readability formulae to mathematics books.

Despite the problems of using readability formulae with ME, it remains important to assess whether a particular piece of ME writing might be 'easy' or 'difficult' or 'about right' for a particular child. Making such an assessment is not a simple matter. To arrive at an informed judgement, it is necessary to look closely at the styles of writing used in ME, and the ways in which children respond to what they read.

Outline of the book

We start by examining the purposes for which authors use mathematical writing (Chapter 2); these purposes include teaching, practice, revision and testing. This leads into a classification of different types of mathematical text. In Chapter 3 we give an account of how the vocabulary and syntax of a mathematical passage affect its readability. In Chapters 4 and 5 we discuss the non-verbal elements of ME; symbols are discussed in Chapter 4, and the graphic language of diagrams, pictures and illustrations in Chapter 5. Words, symbols and pictures make up the detailed level at which text must be read. There are also global features of text which affect its readability: the text must express its intended meaning in an intelligible and complete way. We discuss this problem in Chapter 6.

On the basis of the identification of features of mathematical text, we are able in Chapter 7 to look at the development of readability formulae and their application to mathematical writing. At a very different level, the layout of the material on the page can itself affect readability. This aspect of text is described in Chapter 8.

Chapter 9 brings together many of the ideas presented in earlier chapters, and reports on ways in which we have used these ideas to investigate whether it is possible to write text which children can read more easily.

The problems of reading in mathematics extend beyond matching the reader to the text. How should a teacher best use a particular text in the classroom? How can an author – and most authors are teachers who write worksheets and workcards for their own pupils – write in such a way that the text does not produce undue reading difficulties? How should symbolism best be incorporated into the text? Can the teacher help children to read ME with greater understanding? We discuss these matters in Chapters 10 to 13, and identify three broad directions.

First, a basic concern is the improvement of the written materials used in mathematics. Both the authors of commercially produced books, and teachers who write their own workcards and worksheets, can work to improve their writing. It is often suggested that written materials should not only be 'easy' to read, but that they should be designed to help pupils to become more skilled at reading. For instance, the Cockcroft Report (1982) states in paragraph 311:

> The policy of trying to avoid reading difficulties by preparing work cards in which the use of language is minimised or avoided altogether should not be adopted. Instead the necessary language skills should be developed through discussion and explanation and by encouraging children to talk and write about the investigations which they have undertaken.

Texts can themselves contribute to the development of pupils' reading ability, and this possibility adds a dimension to the task of improving written materials. In Chapter 10, we make a number of suggestions for ways in which teacher-authors can make the text they write more readable.

In Chapter 11, we explore a second aspect: the use of written materials in the classroom. There is an interaction between the style of a written text, the way the pupil responds to the text, and the way the teacher uses written materials in the mathematics lesson. This interaction suggests that the reading of ME should not be examined in isolation from the consideration of classroom teaching styles.

The problem discussed in Chapter 12 is that of teaching children to improve their skills in reading ME. To some extent this issue is concerned with the way texts are written and the ways they are used by teachers. However, explicitly helping children to read ME could initiate a significant change in the content and style of many mathematics lessons. We discuss ways in which the skills needed to read ME may be developed in children.

Finally, Chapter 13 indicates some research areas which might support a growing concern to improve the written communication that is an important part of the teaching of mathematics.

To tackle the challenge of improving the use of written materials in mathematics teaching, it is essential to become aware of the ways in which the printed page communicates with the reader. We hope that the ideas in this book will provide a basis from which teachers can develop skills in helping children to read mathematics.

2 Characteristics of mathematical writing

Introduction

One of the main differences between mathematical text written for children and text used in other subjects is the variety of purposes and types of writing which are often found in mathematics text, even within a single page. In this chapter we identify the different purposes for which mathematical text is used, and discuss and illustrate the types of writing which are used to carry out these purposes; we also analyse a passage of text.

The purposes of mathematical text

In many cases the purpose of a piece of text is clear in the author's mind, but that purpose may not be explicit in the text, and in reading a particular passage, the pupil needs to work out for himself what the author intends by it. Authors and teachers will find that in designing and using a particular passage of text, it is helpful to have defined the intended purpose of the passage.

The list which follows reflects the purposes of current mathematics textbooks. The classification is rather broad, and ignores a number of minor items. Mathematical goals can be briefly summarised as the acquisition of:

- concepts
- principles
- skills
- problem-solving strategies

Teachers and authors also hope that pupils will develop in other ways as part of their mathematical education, but the above list represents the types of goal for which most texts aim. Within these goals, a particular passage of written material may be intended to:

1. *teach* concepts, principles, skills and problem-solving strategies;
2. give *practice* in the use of concepts, principles, skills and problem-solving strategies;
3. provide *revision* of 1 and 2 above;
4. *test* the acquisition of concepts, principles, skills and problem-solving strategies.

In addition, texts also attempt to:

5. develop mathematical *language*, for instance by broadening the pupils' mathematical vocabulary and their skill in the presentation of mathematics in a written form.

In covering the *teaching* phase, particularly the teaching of concepts and principles, many texts give instructions to the pupil to carry out some task. The results of the pupils' work are then used as the basis for identifying the concept or justifying the principle. By this strategy, the author avoids a verbal explanation, which it might not be easy for the pupils to comprehend; he also ensures that learning is active, which conforms well with some philosophies of teaching. However, it is very necessary that the author should devise activities that are suitable and should give instructions that are clear and concise, in order that pupils can in fact attain the concept or learn the principle that the author intended they should.

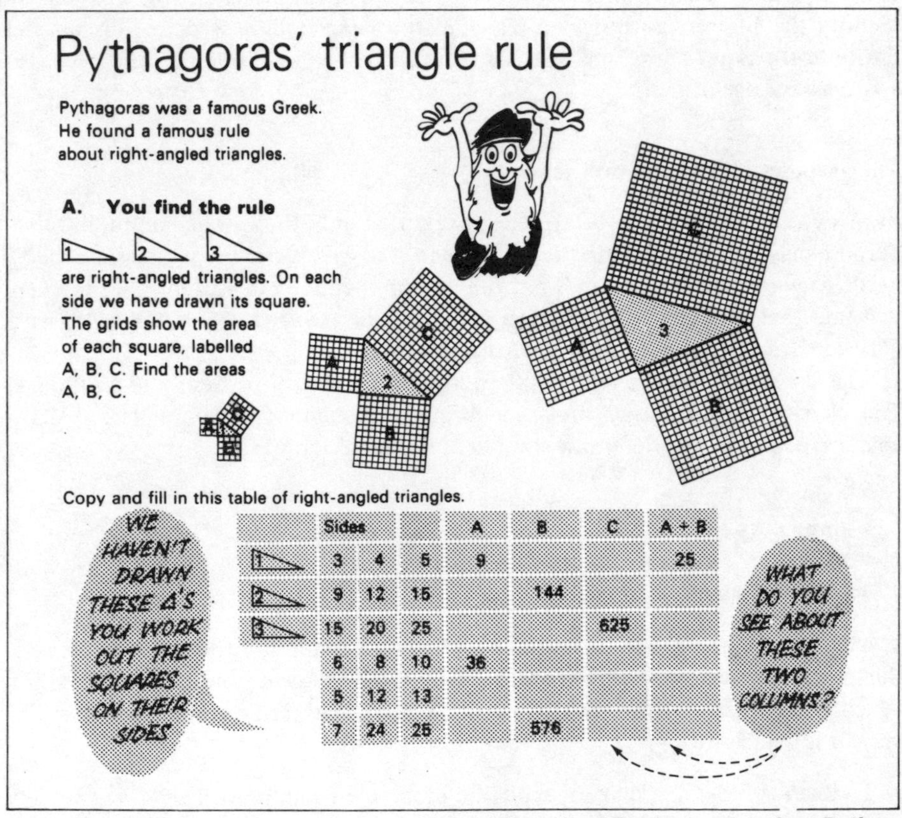

Pythagoras' triangle rule

Pythagoras was a famous Greek.
He found a famous rule
about right-angled triangles.

A. You find the rule

1 2 3 are right-angled triangles. On each side we have drawn its square. The grids show the area of each square, labelled A, B, C. Find the areas A, B, C.

Copy and fill in this table of right-angled triangles.

	Sides			A	B	C	A + B
1	3	4	5	9			25
2	9	12	15		144		
3	15	20	25			625	
	6	8	10	36			
	5	12	13				
	7	24	25		576		

Figure 2.1 M. Holt and A. Rothery (1979).

We give two examples of ways in which authors deal with the teaching phase; in Figure 2.1, the principle of Pythagoras' theorem is introduced by encouraging pupils to notice the pattern in the data which they themselves have produced. In

Figure 2.2, on the other hand, the concept of 'square root' is introduced by straight explanation supported by examples. In this example, the exercises are intended to give practice, and to enable the pupils to find out whether they have grasped the concept.

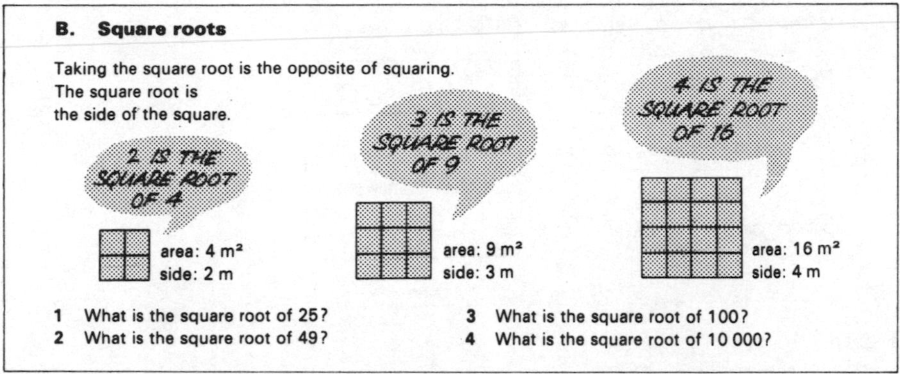
B. Square roots

Taking the square root is the opposite of squaring.
The square root is
the side of the square.

1 What is the square root of 25?
2 What is the square root of 49?
3 What is the square root of 100?
4 What is the square root of 10 000?

Figure 2.2 M. Holt and A. Rothery (1979).

Skills and problem-solving strategies are usually taught by means of worked examples. In Figure 2.3 the skill of using square root tables is treated in this way.

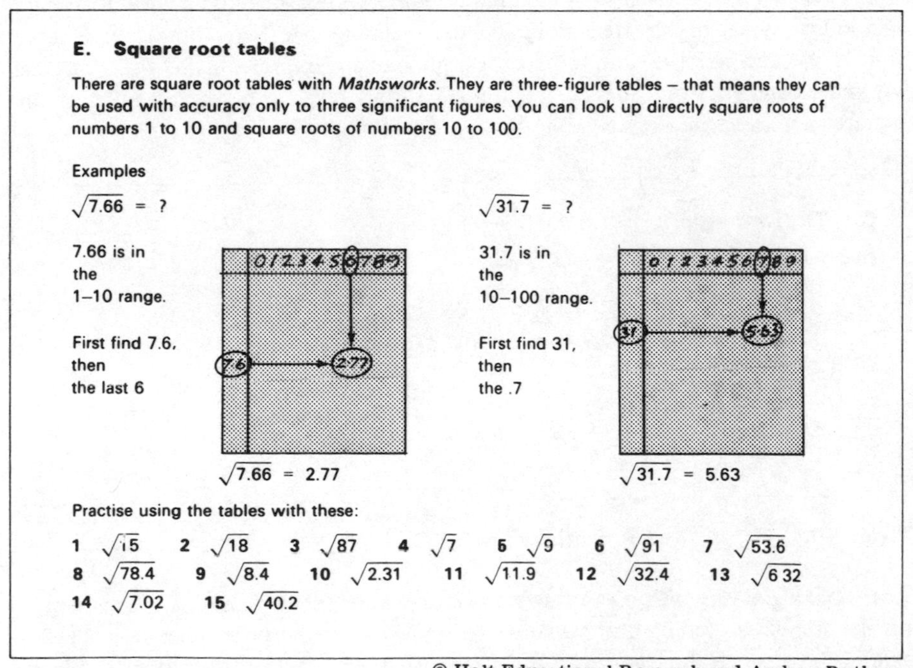
E. Square root tables

There are square root tables with *Mathsworks*. They are three-figure tables – that means they can be used with accuracy only to three significant figures. You can look up directly square roots of numbers 1 to 10 and square roots of numbers 10 to 100.

Examples

$\sqrt{7.66}$ = ?

7.66 is in the 1–10 range.

First find 7.6, then the last 6

$\sqrt{7.66}$ = 2.77

$\sqrt{31.7}$ = ?

31.7 is in the 10–100 range.

First find 31, then the .7

$\sqrt{31.7}$ = 5.63

Practise using the tables with these:

1 $\sqrt{15}$ **2** $\sqrt{18}$ **3** $\sqrt{87}$ **4** $\sqrt{7}$ **5** $\sqrt{9}$ **6** $\sqrt{91}$ **7** $\sqrt{53.6}$
8 $\sqrt{78.4}$ **9** $\sqrt{8.4}$ **10** $\sqrt{2.31}$ **11** $\sqrt{11.9}$ **12** $\sqrt{32.4}$ **13** $\sqrt{6\ 32}$
14 $\sqrt{7.02}$ **15** $\sqrt{40.2}$

Figure 2.3 M. Holt and A. Rothery (1979).

Write-in worksheets of the type shown in Figure 2.4 are often employed to ensure that pupils work actively, rather than passively looking at the author's worked example.

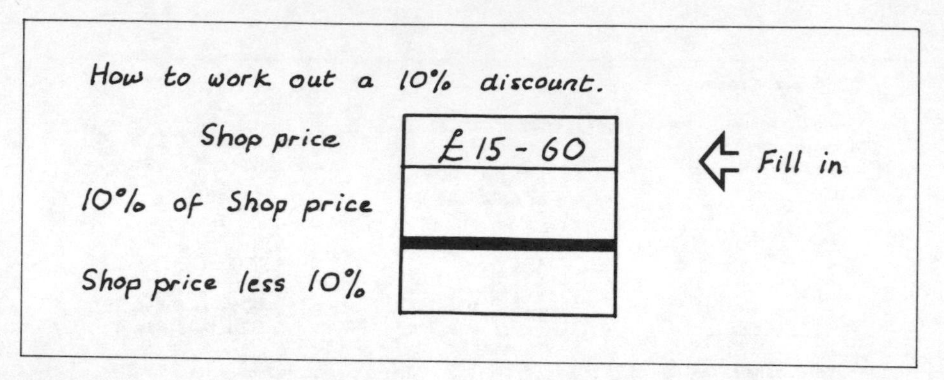

Figure 2.4

The *practice* and *testing* phases are usually carried out through set exercises; indeed, many mathematical texts consist largely or wholly of exercises for practice and testing. However, exercises are also often used as preparatory work for the teaching of a concept or principle, since they can be organised to provide appropriate data to reveal the concept or principle. Figure 2.1 is an example of this strategy.

Language development, such as the building up of mathematical vocabulary and symbolism, is usually treated alongside the teaching of other points: pupils learn how to use and write down mathematical phrases and expressions by following the worked examples. For instance, Figure 2.5 shows revision of the meaning of the 'square root' sign; the exercises reinforce this meaning.

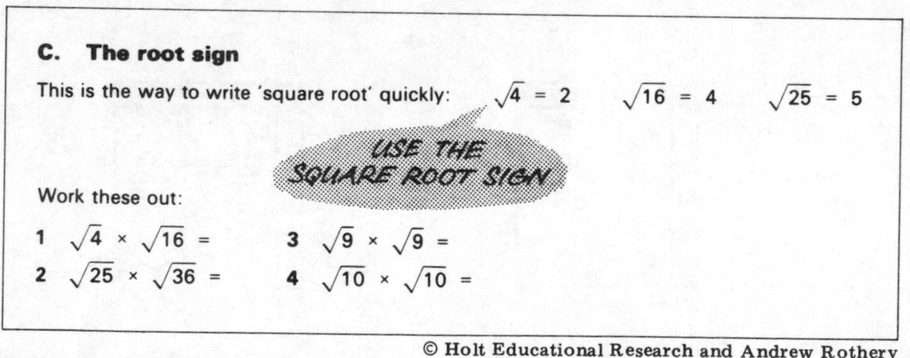

Figure 2.5 M. Holt and A. Rothery (1979).

In looking at the purposes of a few examples of text, we have not commented on the strategies used by authors to try to achieve these purposes; nor have we considered how successful they are. We now turn to a consideration of methods by which authors try to implement their aims.

Classification of types of mathematics text

Mathematical authors use several different types of text in order to carry out their intended purposes, and each type has its own characteristics. Types of text may be classified under the following headings (the labels will be useful later, when we analyse a passage of text):

Expo – *exposition* of concepts and methods, including explanations of vocabulary, notation, and rules; summaries are included in this category.

Instr – *instructions* to the reader to write, draw or do.

Exer – *examples and exercises* for the reader to work on; often these are 'routine' problems involving symbols, but they also include word problems, non-routine problems and investigations.

Periph – *peripheral writing,* such as introductory remarks, meta-exposition (writing about the exposition), 'jollying the reader along', giving clues, etc.

Sig – *signals,* e.g. headings, letters, numbers, boxes, logos.

These categories provide a crude system of analysis. Each category invites a particular kind of response from the reader. The *exposition* is to be read and digested, but not necessarily acted upon immediately. *Instructions* require the reader to move outside the context of the text and to carry out the tasks described. *Examples and exercises* ask for the solution of simple or complex problems; the reader must puzzle out what to do and how to do it, and then carry out his plan and find an answer. *Peripheral writing* can be read in a relatively passive way; in contrast to exposition it does not require full attention and concentration as it does not contain crucial information; nor is it important to remember its content. *Signals* are not 'read' in the conventional sense although they can convey a good deal of information. Their purpose is not explicitly instructional – they are there to guide the reader through the page, helping to identify and clarify the different parts of the text.

There are two very common types of writing which serve many different purposes in mathematical text, and these defy neat classification. One of them is the *question.* Sometimes a question can be classified as an exercise – 'What is the area of a rectangle 7 cm long and 5 cm wide?' Sometimes, especially in the type of text which is written as a dialogue with the reader, questions form part of the exposition. For example, 'What do you expect the next number in the sequence 1, 1, 2, 3, 5, 8, 13, . . . to be?' may be part of an exposition of the Fibonacci sequence. If a pupil understands what is going on in both these examples, then the expected answer to the question is clear to him, whereas it may present problems to a less knowledgeable reader. However, some questions asked in this type of text are not what they appear to be. 'Can you think of any other types of triangle?' is a typical example of this last type of question. To address another question of the same type to our own readers – 'Are you meant to reply, "Yes, I can think of some other types of triangle," or are you meant to get out your pencil and make a long list of all the types of triangle that come into your mind?'

In fact, 'Can you think of any other types of triangle?' may well not be a question which expects an answer. It is much more likely to be a signal that the author is going to discuss other types of triangle in the next passage, and would like the reader to have some of the possibilities in his mind. We shall call questions of this type *rhetorical questions*. Pupils learn that they are expected to attempt to answer all the questions in the mathematics text they are reading; consequently they find rhetorical questions very confusing because it is not clear what response they are expected to make. However, many authors seem to see rhetorical questions as useful signals that they are about to introduce a new idea.

Ormell, in the *Mathematics Applicable* project, made explicit use of rhetorical questions, as the two extracts from the book on polynomial models shown in Figure 2.6 indicate. Extract (a) is from the prefatory chapter 'How to Use this Book', and extract (b) is the first few sentences of Chapter 1.

(a)

◎ indicates a rhetorical question. These questions are not intended to be answered immediately. They are posed in order to clarify mathematical challenges or to introduce discussions of the *purpose* of the current work.

(b)

Forming Linear Models

We begin by looking at the nature of a *linear model.*

◎ What is meant by a 'linear' model?

◎ To what kinds of situation do linear models apply?

Figure 2.6 C. Ormell (1978).

The second commonly used type of text which is not easily classified is the *worked example.* Some worked examples are easily identified as exposition; their format tells the reader how to carry out a particular skill, such as looking up a number in a table of square roots. However, authors often invite the reader to participate in the working of the example, and a response is often essential if the pupil is to follow the text. Hence this worked example becomes a 'guided' exercise and the boundary between exposition and exercise is blurred. Nevertheless, since the major purpose is the same as that of expository text, these worked examples can also be classified as exposition.

Exposition

In reading mathematical exposition the reader often needs to work out steps in the argument for himself, so it is essential for him to use pencil and paper as he reads, rather than passively reading the text as he would a novel. In text for children, some exposition is not distinguished from exercises – indeed, exposition is often presented as exercises in order to force the reader into an active mode of reading. An example is shown in Figure 2.7.

1. PATTERNS AMONG NUMBERS

(*a*)

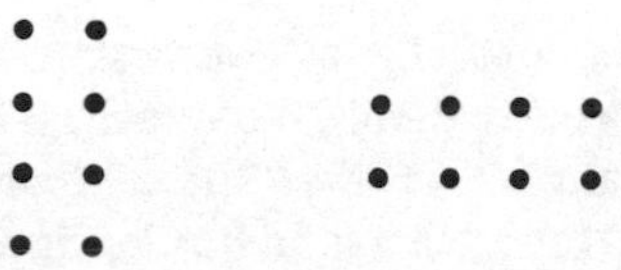

These arrangements of dots both represent the rectangle number 8. They show that

$$4 \times 2 = 2 \times 4.$$

Are the following correct

(i) $9 \times 17 = 17 \times 9$;

(ii) $5 \times 8 = 8 \times 5$;

(iii) $3 \times 405 = 405 \times 3$?

Let *a* and *b* stand for any two members of the set of counting numbers. Is it always true that

$$a \times b = b \times a?$$

Figure 2.7 *SMP Book B* (1979).

Very few examples of exposition by straight explanation are to be found other than in advanced texts. Figure 2.8, p.12, is a piece of exposition from an A-level text; it is general and abstract, and to understand the ideas the reader will probably need not only to work through the exposition with pencil and paper, but also to use the numerical examples which follow, or to invent his own examples to try out the method. In fact, few students at this stage of their education are likely to be enlightened by the general exposition.

Exposition is also used to introduce new vocabulary and notation. This *lexical familiarisation* can be achieved in several ways. Technical vocabulary may be woven into a sentence (as we wove 'lexical familiarisation' into the previous sentence, in the hope that the context and the reader's background would make the meaning

First order linear equations

17.5. A differential equation is *linear in y* if it is of the form

$$\frac{d^n y}{dx^n}+P_1\frac{d^{n-1}y}{dx^{n-1}}+P_2\frac{d^{n-2}y}{dx^{n-2}}+\ldots+P_{n-1}\frac{dy}{dx}+P_n y = Q,$$

where $P_1, P_2, \ldots, P_n$, Q are functions of x, or constants; it is of the nth order.

Thus a *first order linear equation* is of the form

$$\frac{dy}{dx}+Py = Q,$$

where P, Q are functions of x or constants. This type of differential equation deserves special attention because an integrating factor, when required and if obtainable, is of a standard form.

Let us assume that the general first order linear equation given above can be made into an exact equation by using the integrating factor R, a function of x. If this is so,

$$R\frac{dy}{dx}+RPy = RQ \qquad (1)$$

is an exact equation, and it is apparent from the first term that the L.H.S. of (1) is $\frac{d}{dx}(Ry) = R\frac{dy}{dx}+y\frac{dR}{dx}$. Thus (1) may also be written

$$R\frac{dy}{dx}+y\frac{dR}{dx} = RQ. \qquad (2)$$

Equating the second terms on the L.H.S. of (1) and (2),

$$y\frac{dR}{dx} = RPy,$$

$$\therefore \frac{dR}{dx} = RP.$$

Separating the variables,

$$\int\frac{1}{R}\,dR = \int P\,dx,$$

$$\therefore \log_e R = \int P\,dx,$$

$$\therefore R = e^{\int P\,dx}.$$

Thus the required **integrating factor is $e^{\int P\,dx}$**. The initial assumption that an integrating factor exists is therefore justified provided that it is possible to find $\int P\,dx$.

Figure 2.8 J.K. Backhouse, S.P.T. Houldsworth and B.E.D. Cooper, *Pure Mathematics*, Longman, 1963.

clear). The example in Figure 2.9 uses this method to introduce the idea and vocabulary of the Least Common Multiple within the context of a particular worked example; this method is also based on the hope that the pupil will generalise the idea from the single worked example so that he can apply it to other examples.

To compare two fractions, say $\frac{5}{12}$ and $\frac{7}{18}$, we must change them so that they both have a common denominator (bottom part). This denominator is a multiple of 12 and 18, and should be as small as possible, to save work. We call it the LOWEST COMMON MULTIPLE – LCM for short. The highest factor common to 12 and 18 is called the HIGHEST COMMON FACTOR – HCF.

Figure 2.9

For older readers, and sometimes for younger ones too, new vocabulary and ideas may be introduced by means of definitions. Definitions are often signalled by the words 'is called' or 'we say', as in Figure 2.10, which introduces the vocabulary of an inequation and its graph.

With the verb 'is greater than', or 'is less than', the sentence is called an *inequation*.

$x > 3$ and $x + 2 < 6$ are inequations.

This picture of the solution set on the number line is called the *graph* of the solution set of the inequation $x > 3$, which we shorten to 'the graph of $x > 3$'.

Figure 2.10 Scottish Mathematics Group, 1971.

The sets are proportional

⇔ the ratios of corresponding numbers are equal
⇔ the ordered pairs (d, p) are equivalent
⇔ the ratio d:p is *constant* (fixed)

Figure 2.11

The passage in Figure 2.11 uses a strategy which is intermediate between using a definition to achieve lexical familiarisation and achieving the result by incorporating the word in a sentence. However, there is little clue that the passage is intended to introduce the word *proportional,* while the three statements said to be equivalent to it may not seem familiar and equivalent to the pupil, and the emphasis on *constant* suggests that this may be the subject of the definition.

Asymptotes. Consider the curve $y = \dfrac{1}{x-1}$.

As x approaches the value 1, y approaches either $+\infty$ or $-\infty$, according as x is greater than or less than 1. We say that the line $x = 1$ touches the curve at infinity.

The line $x = 1$ is called an *asymptote* of the curve.

Asymptotes parallel to the axes can be found by solving the equation of the curve for y and for x.

Figure 2.12 R.I. Porter (1951).

The passage in Figure 2.12 is intended to familiarise the idea of an asymptote. Perhaps the student is intended to extract from this passage the definition that 'a line which touches a curve at infinity is called an asymptote', whatever he may take this to mean.

In the example of Figure 2.14, the temporary definitions of the informal wording 'blue shift' and 'green shift' receive exactly the same emphasis as will their permanent replacement, when the pupil is told later in the chapter that 'in all these situations we could use the set of directed numbers and we shall now use this set to describe shifts to the right and to the left' (*SMP Book C,* p. 28).

A definition may be implied within some other piece of text; for example, at the end of the exercise about blue and green shifts which follows the extract in Figure 2.14, we find the question: 'What name could you give the shift number which leaves the counter where it is? Is this a blue shift or a green shift?'. This question implies the definition: 'There is a shift number called zero'.

Notation, as well as vocabulary, needs to be defined in mathematical text, and the same techniques of lexical familiarisation are used. Important standard notation may be introduced within a sentence in both elementary and advanced work. Figure 2.13 shows the introduction of notation for 'shift numbers' in *SMP Book C,*

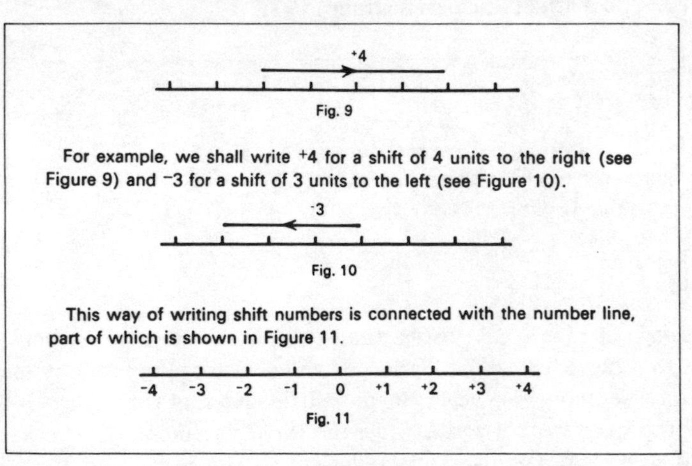

For example, we shall write $^{+}4$ for a shift of 4 units to the right (see Figure 9) and $^{-}3$ for a shift of 3 units to the left (see Figure 10).

This way of writing shift numbers is connected with the number line, part of which is shown in Figure 11.

Figure 2.13 *SMP Book C* (1969).

2. Directed numbers

In this chapter we shall be looking at shift numbers. They are called *shift numbers* because they tell us to shift or move along a certain number of places.

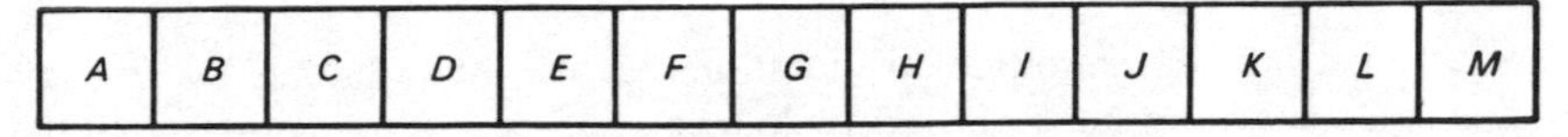

Fig. 1

1. SINGLE SHIFTS

Suppose you place a counter on the square marked *G* in Figure 1. You could move this counter either to the right or to the left. So a move of 2 could take the counter either to *I* or to *E*. In order to tell the difference between instructions to move to the right and to the left, moves to the right are given by 'blue shifts' (see Figure 2) and moves to the left by 'green shifts' (see Figure 3).

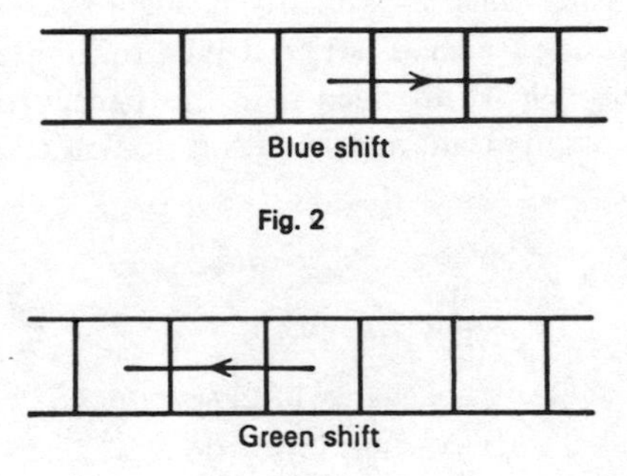

Fig. 2

Green shift

Fig. 3

So starting from *G*, a blue 2 tells you to move your counter 2 squares to the right, to *I*, while a green 2 tells you to move your counter 2 squares to the left, to *E*.

(*a*) If you started at *D*, where would your counter be after the shift 'blue 3'?

(*b*) If you started at *H*, where would your counter be after the shift 'green 4'?

Figure 2.14 *SMP Book C* (1969).

A more sophisticated example is to take any set X and let $\mathbb{P}(X)$ be the set of all subsets of X. This is called the *power set of X* and satisfies the property:

$$Y \in \mathbb{P}(X) \quad \text{if and only if} \quad Y \subseteq X.$$

Figure 2.15 I. Stewart and D. Tall (1977).

within a particular example of their use; Figure 2.15 is an example from an advanced book of the introduction of both an idea and its notation by their use within a sentence.

Examples of implicit definitions of notation can also be found, as in 'To show which elements belong to a set, curly brackets { } are often written'; having encountered this statement, the reader will later be expected to be able to identify the elements of a set by the curly brackets which enclose them.

In all these examples the reader has to pick out what is important, grasp it and store it in his memory. There are also many less important points which he does not need to remember. If a teacher is not always at hand, then the text should clarify which points are important.

In some styles of exposition found in mathematical text, the statement of *rules* forms an important part of the text; sometimes the word 'rule' signals the importance of what is to follow. Rules are sometimes stated with explanations and sometimes without any justification; they may be labelled as rules, or the reader may be left to infer for himself that a particular statement is a rule. Figure 2.16 gives an example of a rule which is explicitly labelled as a rule, together with an example in which the young reader is left to decide for himself that the statement is a rule; however, some help is provided – in this particular series of texts, it is usual to enclose especially important matter within a coloured box.

(a) The change δT could be given approximately as

$$k\,\delta p + p\,\delta k.$$

As a rule that would give the change in any product, it might be stated as: Multiply each part by the change in the other and add.

(b) To multiply by 10
Move the figures ONE place to the LEFT.
Put a 0 in the units column.

H	T	U		H	T	U
	1	8	× 10 =	1	8	0
	5	9	× 10 =	5	9	0

Figure 2.16 (a) An explicit rule. SMP (1973). (b) An implicit rule. T.R. Goddard and A.W. Grattidge (1969).

The distinction between *rules* and *summaries* is often blurred. A summary is often the statement of a rule which the pupils were intended to discover for themselves through exercises or activities provided earlier in the text. This technique allows the author not to 'give the game away' in the discovery phase.

Instructions

Instructions to write, draw or do something are very common in elementary texts. 'Copy and complete' is an instruction which occurs very often in text for young children, but it does not occur in advanced texts, although the older reader also needs to work through examples and proofs for himself, completing arguments which are not fully stated.

A curious feature of some elementary texts is that there are some instructions that cannot be taken literally; they *cannot* be carried out, although they look like real instructions. For example, an exercise from the page which introduces division in *Maths Adventure 3* is shown in Figure 2.17. If the pupil needs to draw and cut

2 Draw a stick 21 cm long and cut it into:

a 3 equal pieces

b 7 equal pieces

c 2 equal pieces

Record the length of the pieces in each case.

Figure 2.17 J. Stanfield (1971).

up a picture of a 21 cm stick in order to find the size of each one of 3 equal pieces then he will certainly not know how to cut the stick into 7 reasonably equal pieces. Pupils quickly recognise which instructions they should take literally. They also recognise those other instructions which cannot feasibly be carried out, and they know that all that could have been intended was a 'thought experiment', which was a mathematical model of the actual instructions. However, the decisions which pupils make are not always those intended by the author (if he had, in fact, thought how his instructions were to be carried out!).

Exercises

Exercises can be classified in terms of the closeness of their connection with the previous textual matter, as in the following list. In each case, an example of the type is shown in the figure.

Type 1 Exercises *without* extension: these exercises are intended for direct practice of the idea previously introduced (Figure 2.18).

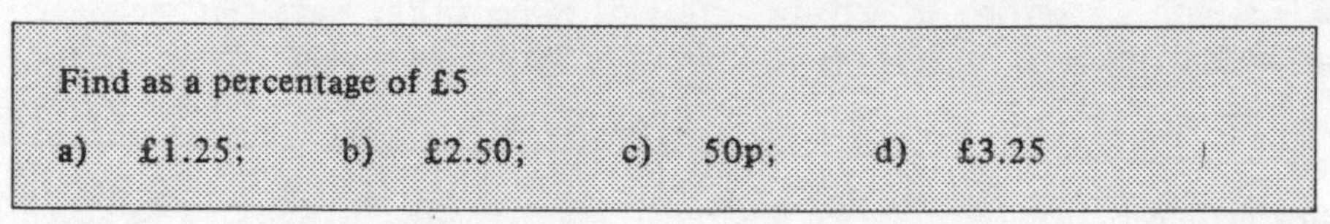

Find as a percentage of £5

a) £1.25; b) £2.50; c) 50p; d) £3.25

Figure 2.18 A Type 1 exercise.

Type 2 Exercises *with* extension: these exercises require the pupil to apply his knowledge to some small extension of the idea presented in the text (Figure 2.19).

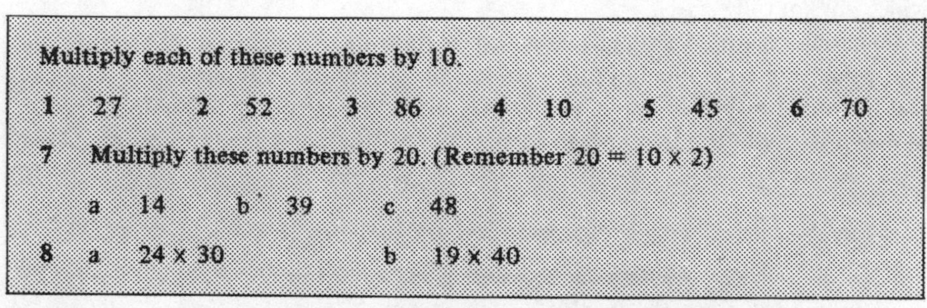

Multiply each of these numbers by 10.

1 27 2 52 3 86 4 10 5 45 6 70

7 Multiply these numbers by 20. (Remember 20 = 10 × 2)

a 14 b 39 c 48

8 a 24 × 30 b 19 × 40

Figure 2.19 A Type 2 exercise. T.R. Goddard and A.W. Grattidge (1969).

Type 3 Word problems: here the process is set within a description in words of a 'real life' situation, in which the pupil is meant to discern the intended process (Figure 2.20).

The mortgage on Jim's house is £3000. How much more interest will he have to pay when the interest he pays to the building society goes up from 12% to 13%?

Figure 2.20 A Type 3 exercise.

Type 4 Picture problems: these are similar to word problems, but diagrams and illustrations are used to convey the problem (Figure 2.21).

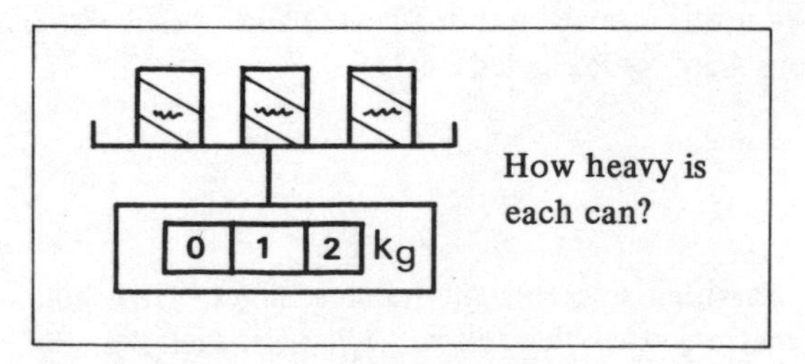

Figure 2.21 A Type 4 exercise

Type 5 Miscellaneous problems and investigations: here there is no necessary connection between the preceding text and the problem; the reader is expected to use any knowledge he has in order to explore the investigation or solve the problem (Figure 2.22).

How many squares are there on a chessboard?

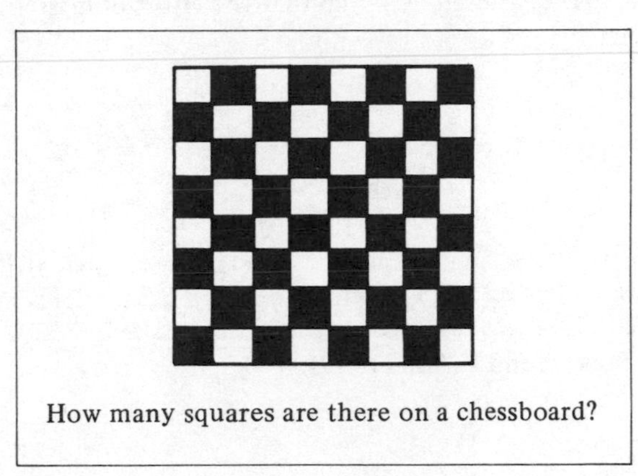

Figure 2.22 A Type 5 problem

There are also some passages which look like exercises, but which are not designed as exercises for individual pupils to carry out; these exercises are intended to be used for class or group discussion and to enable teaching points to be made. The teacher should be on the lookout for examples of this type; sometimes for convenience, they are tucked away at the ends of exercises, and they are sometimes found in the main body of the text. They cannot be recognised from the brief extracts reproduced here; the teacher needs to examine the whole passage. These exercises are intended as discussion points which can be used to draw out new ideas implicit in the passage which has gone before.

Peripheral writing

In mathematical text, as in other types of writing, introductory passages provide links with previous ideas, or give pointers or advance organisers to help the reader to fit what follows into some framework. Examples of a linking passage and a pointer are shown in Figure 2.23.

(a) In the last chapter you learnt how to add fractions:

$\frac{1}{4} + \frac{2}{3} = \frac{3}{12} + \frac{8}{12} = \frac{11}{12}$

(b) Let us take up the working of Questions 3 and 5 and see how it may be used to obtain a rule that can be generally applied to any function.

Figure 2.23 (a) A linking passage. (b) A pointer. SMP (1973).

Authors sometimes use meta-exposition – talking about the exposition – to explain its purpose, to comment on the text, to 'jolly the reader along' by making encouraging remarks, or to give warnings that all is not straightforward. An example is shown in Figure 2.24. Here the authors clearly think that their notation will baffle the reader, so they pause to tell him that the effort of mastering the notation will be worthwhile.

In general, for any set X, we have

$$\cup \mathbb{P}(X) = X$$
$$\cap \mathbb{P}(X) = \emptyset$$

Although this notation may seem a little strange at first, it is extremely economical and it does act as a genuine extension of the usual concepts.

Figure 2.24 I. Stewart and D. Tall (1977).

Signals

Signals are a very important part of mathematical text. Because the writing is often complex, and the layout of the page is not straightforward, signals are often urgently necessary to tell the reader which exercise to do next, to warn him that some statement is particularly important, or indeed to tell him what the passage is about. Headings of chapters and sections may introduce important vocabulary as well as telling the reader the subject of the work (Figure 2.25).

CHAPTER 10 **PROPORTION, PROFIT AND INTEREST**

Figure 2.25

Bold type and italics are used to provide signals which point to definitions and other important statements; in handwritten or typed material, underlining is used for the same purpose (Figure 2.26).

If we divide a cake into four equal parts as in Figure 1, then each of these parts is one quarter ($\frac{1}{4}$) of the whole cake.

If three of the four equal parts are taken we have *three quarters* ($\frac{3}{4}$) of the cake. $\frac{1}{4}$ and $\frac{3}{4}$ are called *fractions*.

In the fraction $\frac{1}{4}$, 1 is called the *numerator* and 4 the *denominator*.

Scottish Mathematics Group (1971).

This shows the equivalence of these fractions, which are really different names for one fraction: $\frac{1}{3}$.

Figure 2.26

Typographical signals such as numbers, letters and roman numerals indicate new sections; boxes, arrowheads, block capitals and colour draw attention to important passages; in more advanced texts, equations which are referred to later may be numbered. All these typographical devices help the reader to find his way around what would often otherwise be a very crowded and jumbled page.

Children need to learn the conventions of the typographical signals used in mathematics. For example, the convention shown in Figure 2.27(a), that the 'stem' of a question should be taken to apply to each of its parts, may cause

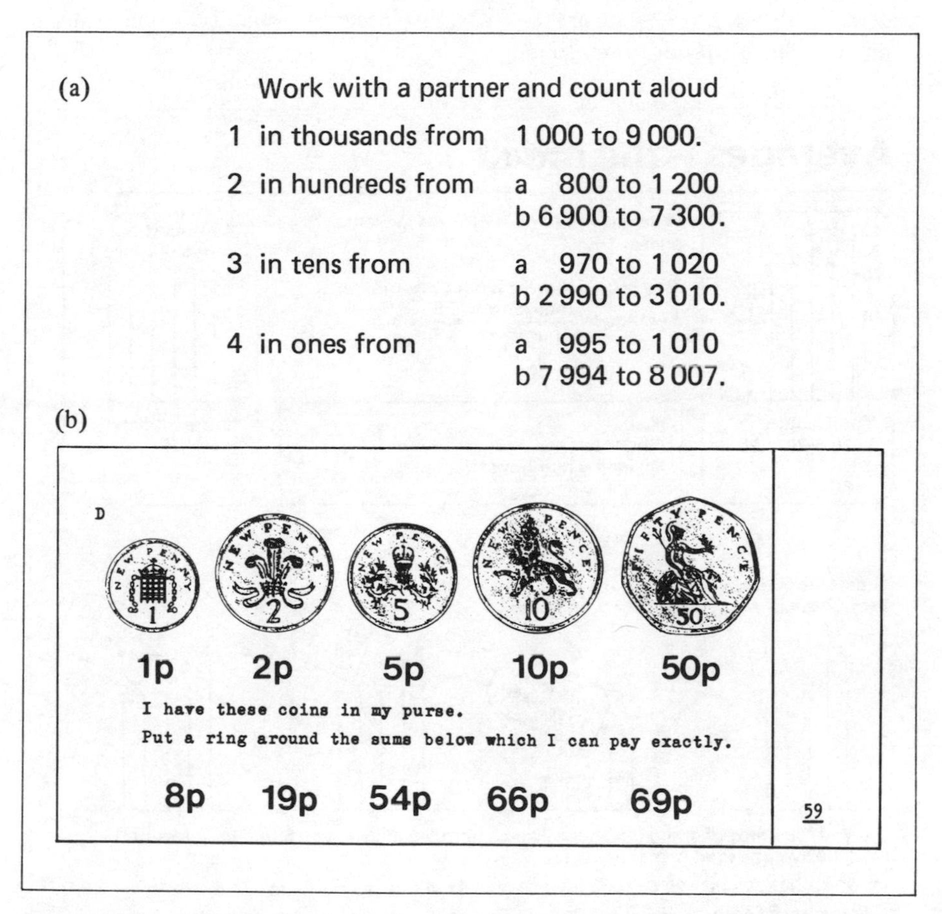

(a)

Work with a partner and count aloud

1 in thousands from 1 000 to 9 000.

2 in hundreds from a 800 to 1 200
b 6 900 to 7 300.

3 in tens from a 970 to 1 020
b 2 990 to 3 010.

4 in ones from a 995 to 1 010
b 7 994 to 8 007.

(b)

D

1p 2p 5p 10p 50p

I have these coins in my purse.
Put a ring around the sums below which I can pay exactly.

8p 19p 54p 66p 69p

59

Figure 2.27 (a) T.R. Goddard and A.W. Grattidge (1969), (b) DES (1978).

children who do not know this convention to give very surprising answers; a well-known example is given in Figure 2.27(b), when only 59 per cent of the 11-year-old children tested during the National Primary Survey (DES, 1978) gave both 8p and 66p in answer to the question. To many others, no doubt, once a coin had been used it was spent and could not be used again.

An example

Figure 2.28 on the following pages gives an example of the various types of text we have discussed. The sample passage is taken from a draft of a text intended for use with pupils of below average attainment in the fourth or fifth year of the secondary school. Figure 2.28(b) contains examples of five categories of text, labelled with the appropriate codes (see opposite page). It also contains two examples of what the pupil must regard as rhetorical questions. The passage contains short sentences and simple vocabulary, and the Fry graph suggests that the text is of a reading level which would be suitable for 9- or 10-year-old children. Whether the intended readers of the passage, who are low-attaining secondary pupils, would find it comprehensible is for our reader to judge.

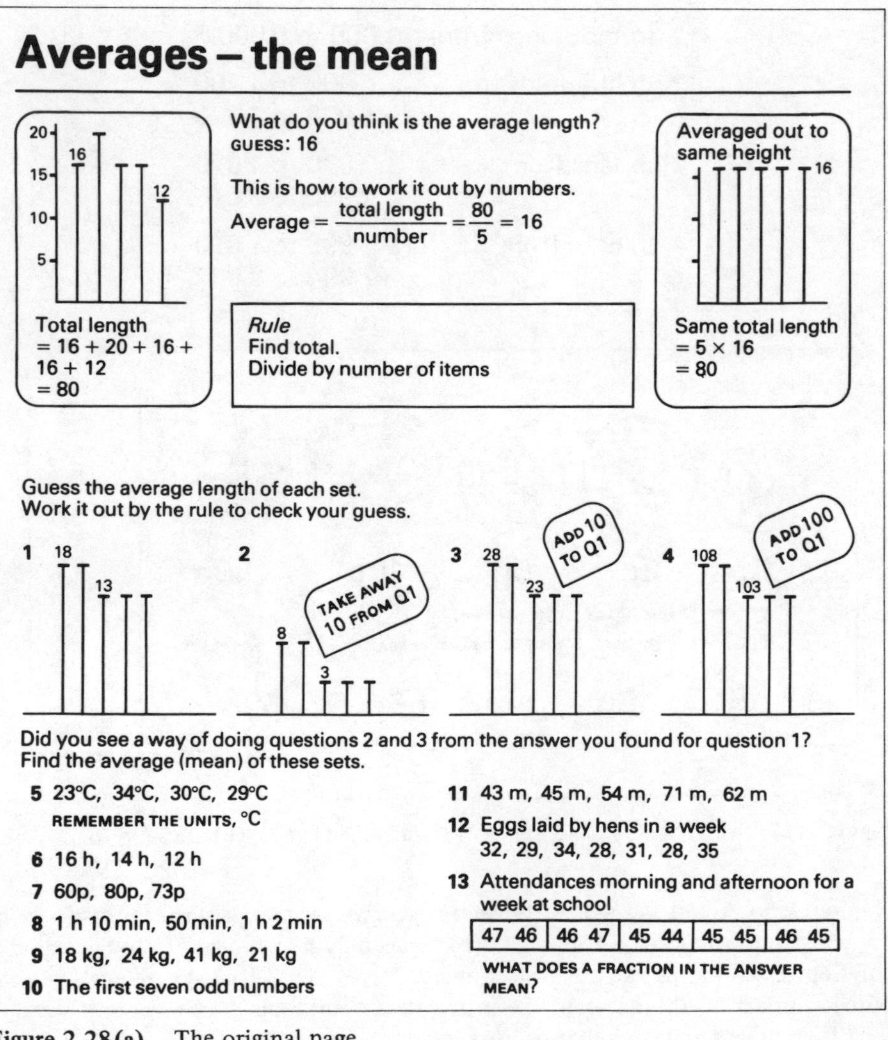

Averages – the mean

Total length = 16 + 20 + 16 + 16 + 12 = 80

What do you think is the average length?
GUESS: 16

This is how to work it out by numbers.

$$\text{Average} = \frac{\text{total length}}{\text{number}} = \frac{80}{5} = 16$$

Rule
Find total.
Divide by number of items

Averaged out to same height

Same total length = 5 × 16 = 80

Guess the average length of each set.
Work it out by the rule to check your guess.

1 18 13

2 8 3 TAKE AWAY 10 FROM Q1

3 28 23 ADD 10 TO Q1

4 108 103 ADD 100 TO Q1

Did you see a way of doing questions 2 and 3 from the answer you found for question 1?
Find the average (mean) of these sets.

5 23°C, 34°C, 30°C, 29°C
REMEMBER THE UNITS, °C

6 16 h, 14 h, 12 h

7 60p, 80p, 73p

8 1 h 10 min, 50 min, 1 h 2 min

9 18 kg, 24 kg, 41 kg, 21 kg

10 The first seven odd numbers

11 43 m, 45 m, 54 m, 71 m, 62 m

12 Eggs laid by hens in a week
32, 29, 34, 28, 31, 28, 35

13 Attendances morning and afternoon for a week at school

47 46	46 47	45 44	45 45	46 45

WHAT DOES A FRACTION IN THE ANSWER MEAN?

Figure 2.28 (a) The original page.

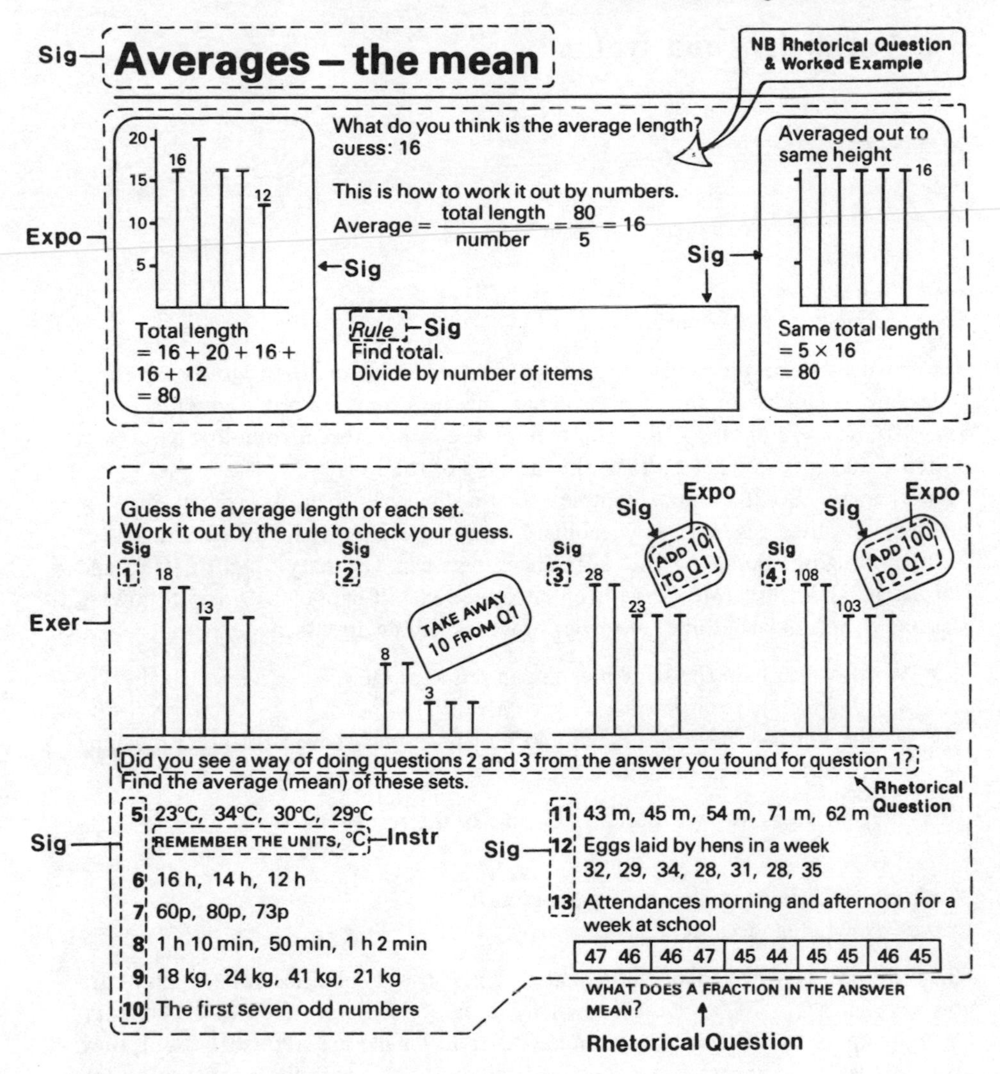

Figure 2.28 (b) Categories of text marked.

3 Vocabulary and syntax

The vocabulary of mathematics

'Hard' words cause problems for the reader. Generally, the hard words are those which are unfamiliar to the reader, perhaps because they are not frequently used. The difficulty of a word is often softened by its context; the surrounding passage or sentence may give valuable clues to the meaning of the word.

Mathematical text is more complex than ordinary English text, partly because mathematics uses a technical vocabulary which overlaps with the vocabulary of ordinary English. Some of the differences between Ordinary English (OE) and Mathematical English (ME) have been encountered in Chapter 2. In the following sections on vocabulary, three categories of words will be discussed.

- Words which have the same meaning in ME as in OE.
- Words which have a meaning only in ME.
- Words which occur in both OE and ME, but which have a different meaning in ME from their meaning in OE.

Each category poses its own special problems for the reader.

Words which have the same meaning in ME as in OE

Examples: cat, dog, because, it, taxi, shelves, climb

When OE words are used in mathematics texts, the difficulties for the pupil are not, on the face of it, worse than when the same words are used in other texts. On the whole, the OE words are familiar to children and are frequently used; they provide oases of easy reading amidst the symbolic and technical content of the mathematics text. Earp and Tanner (1980) investigated the words used in an American sixth-grade textbook. The book used 716 words as a basis, and only 197 of these were used in a technical mathematical way (e.g. average, milliliter, scale, thousandths). Fifty pupils were interviewed to assess their comprehension of the vocabulary of the text; it was found that common non-mathematical words had a comprehension accuracy of 98 per cent compared with only 50 per cent for the mathematical words.

However, even simple words are sometimes used in mathematics in a way that detracts from their simplicity. Earp (1971) remarked that the context provided by a mathematical passage is often less rich than the context of an OE passage: the number of context clues to the meaning of an unknown word or phrase is smaller, and it might therefore be more difficult to read than the same word or phrase in an

OE passage. As an example we take a question from a 'home-made' worksheet which was intended to lead up to more formal work on fractions (Figure 3.1).

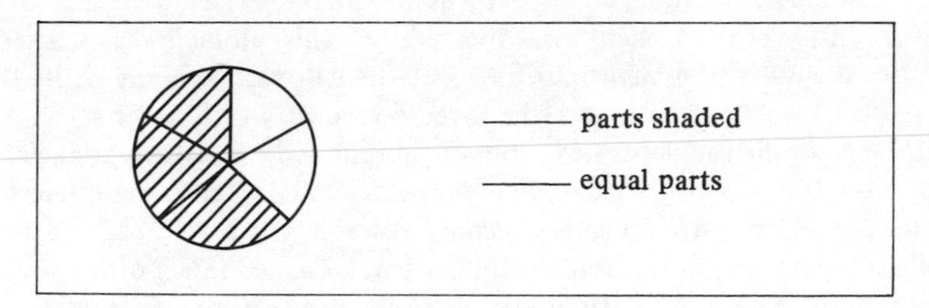

Figure 3.1

Pupils had to complete the phrases, both of which were apparently written in OE. Most pupils correctly completed '3 parts shaded', but several alternatives were suggested for the second phrase; '2' and '3' were given as well as '5'. Evidently the message contained in the words *equal parts* was unclear. Indeed, '— parts altogether' would be clearer, but might still be open to misinterpretation. Such ambiguity in an OE text may not prevent pupils from understanding it, but in mathematics exact comprehension is often crucial, and there may not be enough context clues for exact comprehension. The problem is not that the pupils were incapable of counting the number of parts into which the circle had been divided, but that they misinterpreted what was asked of them. Some expansion of the text (as in Figure 3.2) would have provided a richer context.

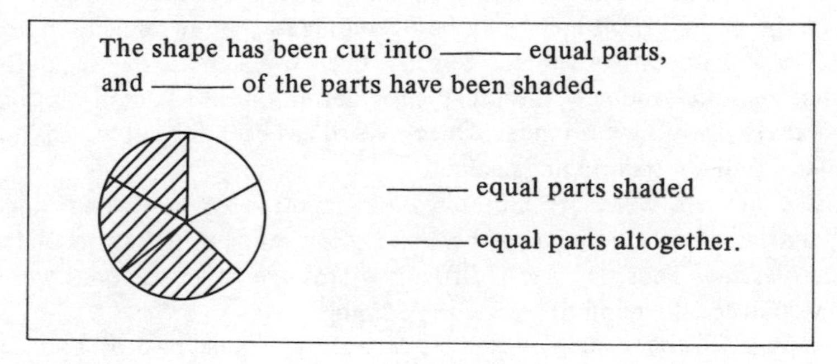

Figure 3.2

Words which have a meaning only in ME
Examples: hypotenuse, parallelogram, coefficient

Pupils meet these words only in a mathematical context and their meanings must be learnt either from the teacher or the mathematics book. Mathematical words are unlikely to be used at home or in the child's everyday speech, and so they cause reading difficulties simply because they are rare in the child's experience. A further difficulty is that many mathematical words have Greek origins. For example, the word *isosceles,* which derives from the Greek words *isos* (equal) and *skelos* (leg), proclaims its meaning clearly to someone who understands Greek, but to most children nowadays, the roots of these words are unfamiliar, and they cannot associate them with words in their own vocabulary. This problem has existed since Greek geometry was translated into English. When Robert Recorde wrote the first geometry book in English in 1551 he invented words which had English roots in order to express the geometrical vocabulary he required, so that the triangles which we call (after the Greek) 'equilateral', 'isosceles' and 'scalene' were described by Recorde as *threelike*, *twolike* and *nonelike* (Howson, 1982). Recorde's attempt at an anglicised mathematical vocabulary did not catch on, and much of the vocabulary of ME still has Greek or Latin roots. Consequently, it has unfamiliar spelling patterns which make initial recognition and verbalisation of many ME words difficult for pupils.

Not only are mathematical words intrinsically unfamiliar and difficult, but they may even be difficult for children to learn in the mathematics lessons. Austin and Howson (1979) stress the point that

> In England the tradition is to rely on the formal definition in higher education and informal or ostensive definitions in primary and secondary schools.

The implication of this statement is not a plea for formal definitions in the primary classroom, but a recognition of the fact that, in speech, both teacher and pupils often use informal language, leaving the formal language to be read in the textbooks. Thus there is a danger of pupils not having the language experience and the clear explanations of mathematical words which they need to support their reading of the text; this problem may be overcome by a more explicit approach to technical vocabulary in mathematics lessons. Even when the text gives definitions it may not reinforce the new vocabulary, for definitions are seldom repeated and children's texts rarely have an index. Once a word has been forgotten, it is not easy for a child to find out its meaning unaided.

To make matters worse, technical words are often of key significance in a passage, and failure to comprehend them can often result in failure to understand the whole passage. Thus, the most difficult words are often the ones which it is most important for the pupil to read with meaning.

One response to the reading difficulties caused by technical words is for teachers and authors to avoid them. For example, words such as *minuend* and *subtrahend* have more or less disappeared from use in textbooks. However, *numerator* and *denominator* are still used, although many texts replace them by phrases such as *number on top* and *bottom number of the fraction*. Omitting all technical words is a short-term policy which makes text easier to read, but it may bring long-term disadvantage to the pupil. Many technical terms have an essential place in mathematics; children cannot proceed without knowing them. Pupils will have even greater difficulty in learning words if they never meet them in their reading. So a practice which may seem to be a kindness, may in fact lessen pupils' experience of essential vocabulary, and so may work against the pupils's future comprehension of mathematics.

It is desirable to avoid the unnecessary use of unimportant technical vocabulary. Rather than omitting *important* vocabulary, however, a better response by the author is to help the pupil to read and become familiar with the words he needs. Strategies such as repeated definitions, marginal comments, reminders, stronger context clues, a glossary and an index, may be useful approaches. A systematic structured approach to teaching and using technical words is not found in many texts at present, and introducing the vocabulary of ME remains very much a job for the teacher.

Words which have different meanings in ME and in OE
Examples: difference, product, parallel, odd, mean, value

Words in this category are very common in ME. The main cause of their reading difficulty is the confusion which can arise when a word has a variety of meanings, and when the required meaning has to be inferred from the context.

Some words have a mathematical meaning which is unrelated to their everyday usage. The child who answered the question 'What is the difference between 7 and 10?' with 'One is odd and the other is even' would no doubt be mystified to learn that the officially approved answer is '3'. The mathematical meaning of the word *product* is only remotely linked with its ordinary meaning, but in the case of *gradient* the mathematical meaning does have a connection with the everyday meaning.

Thus, there are two types of words which have different meanings in ME and in OE. There are some words, such as *product* and *difference*, where the connection between the two meanings is either non-existent or so slight as not to be apparent to the reader. On the other hand, there are some words where the meaning in ME is similar to the meaning in OE, but where the mathematical word has a more refined or specialised meaning. Such words include:

similar circular divide average reflection remainder

For example, the two shapes in Figure 3.3 look *similar* in the general everyday sense of the word, because they are both right-angled triangles. But the more specialised mathematical meaning of the phrase *similar triangles* requires that the corresponding angles of each triangle should equal one another. The triangles in Figure 3.3 are not similar in the mathematical sense. In contrast, the triangles in

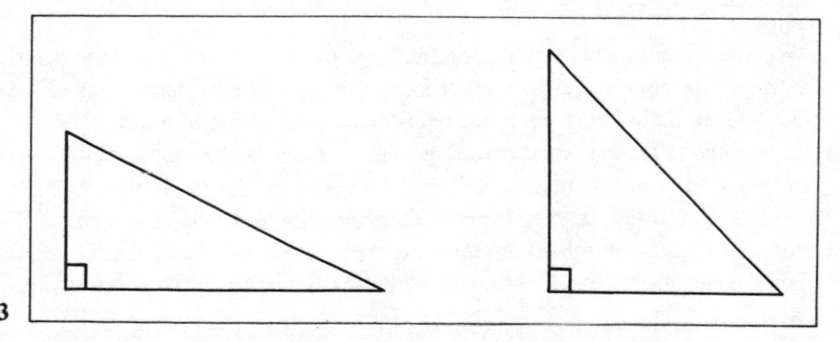

Figure 3.3

Figure 3.4 are *similar* in a technical mathematical sense. A pupil who reads a passage discussing these two triangles could see the word 'similar' and interpret it

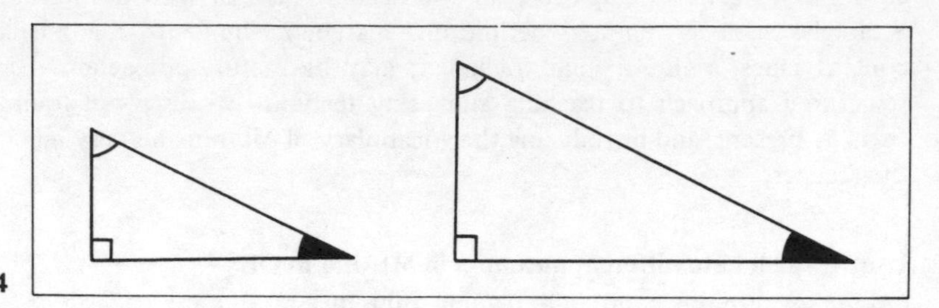

Figure 3.4

in the broader sense, and so his reading may not help him to understand the technical mathematical meaning of 'similar'. Austin and Howson (1979) note that a child will be forming his own internalised definitions of words as his experience progresses, but

> if his usage, say of *similar*, embraces that of the more limited mathematical one, then no apparent contradiction will occur when he *receives* information. It is only when he *transmits* it that inconsistencies may become apparent.

In other words, the pupil may make some sense of his mathematics lessons, although he is only thinking in terms of the colloquial meaning without fully grasping the extra subtlety of the full mathematical meaning.

One way in which mathematics takes over ordinary words is by means of giving specialised meanings to short phrases, such as:

simple interest   pie chart   square root   significant figure

In each case, the individual words have a meaning in OE, but the combined phrase only has a meaning in ME.

Children may discover the alternative meanings of a word at different stages. Sometimes the mathematical meaning of a word is encountered first and then later the child finds that the word has another meaning. For instance, *parallel* can be found in many mathematical texts intended for primary children who would be unlikely to use the word in its OE sense, as in 'there are parallels between these two stories'. On the other hand, a word such as *difference* may be encountered first in its OE sense, and the ME meaning comes as a surprise, or as a refinement of meaning.

Words which have a mathematical meaning may have several mathematical meanings: number systems, triangles, vector spaces and logarithms all have *bases*, and none of these have much connection with the base of a model or box, or with a base villain. Several dual-meaning words carry a derogatory connotation, and pupils may detect this nuance without realising that it is prejudicing them against the idea. Fractions may be *vulgar* or *improper*, a *mean* is a rather 'underhand' average, a *negative* number feels less good than a *positive* number, and when numbers become *irrational* or *imaginary*, mathematics has apparently entered the realms of nonsense.

In OE as well as ME, words often have more than one meaning; the context tells us which meaning is meant. Mathematics is often weak in its provision of *context clues,* so that the pupil who needs this help gets less of it than he does in other types of reading. For example, a pupil who is unsure about the meaning of 'express' or 'standard form' in the following passage will get little help from the context:

> The nearest star to Earth is about 40 000 000 000 000 kilometres away. This number is easier to write if we express it in standard form: 4×10^{13}.

Although exposition does not contain many context clues, exercises generally contain even fewer. Many exercises are designed largely to test pupils' understanding of the meaning of the vocabulary:

> Express the following in standard form, rounding each off to 3 significant figures:
> (a) 2468 (b) 303030 (c) 234 (d) 13579

Here there is no help from context; the pupil must understand each word exactly in order to cope with the exercise.

An investigation of pupils' difficulties with mathematical words

Otterburn and Nicholson (1976) investigated pupils' understanding of mathematical words used in CSE mathematics courses; they classified the pupils' responses to each word as:

correct demonstrating clear knowledge of what the word means

blank the pupil does not give any indication that he knows what the word means, although he may recognise the word

confused generally muddled comprehension

Some examples of the words and the results of the testing are given in the table. The sample consisted of some 300 pupils.

	Percentage of pupils		
Word	Correct	Blank	Confused
multiply	100	0	0
remainder	92	8	1
reflection	45	51	4
mapping	16	81	3
product	21	59	20
integer	15	76	9

This research makes it clear that there is a serious problem. Many ME words which teachers, authors and examiners expect to be understood seem to be absent from the vocabulary of a large proportion of pupils, or at best, they have only a vague or confused idea of what the words mean.

Nicholson (1977) followed up this work with a second investigation in which he gave context clues for the words. The words were given in a sentence or a group of sentences rather than in isolation. The results of this investigation (with another group of pupils) were:

Word	Percentage of pupils		
	Correct	Blank	Confused
multiply	100	0	0
remainder	98	0	2
reflection	87	1	13
mapping	41	33	27
product	21	5	75
integer	10	19	72

Although the two investigations are not comparable, we again see that pupils have problems with some common ME words. Some words, however, 'improved' in the second investigation. This might be due to the different background of the pupils, or it may be that some words are more susceptible to context clues than are others. Indeed, further research might probe the effect of context clues on the reading of different categories of words. Earp and Tanner (1980) found that pupils could define up to 9 per cent more words by using context clues from the textbook, and this could be improved to 15 per cent by using additional context clues.

Assessing the difficulty of vocabulary

In the tests of readability mentioned in Chapter 1, two measures are suggested as an indication of the difficulty of words: word length and word familiarity.

In many readability formulae, *short words* are regarded as easy and *long words* as difficult. This is only a crude mathematical model of the difficulty of words; it is perhaps explained by the fact that there is a tendency for short words to be more familiar, but it may also be the case that longer words are harder to vocalise and so are less easily recognised by readers. However, in assessing a piece of text, the general guideline is not always very reliable, as there are many examples of unusual short words and there are some very familiar long ones.

Perera (1980) surveyed the value of readability formulae and found that:

> Teachers have to assess the linguistic difficulty of a text in order to provide pupils with reading material at an appropriate level. It is argued that informed judgements by a thoughtful teacher may have advantages over the application of a readability formula.

In this spirit, a teacher who is examining a text, or who is in the process of rewriting a workcard, should not stick slavishly to the 'long words are difficult' principle.

The familiarity of words is perhaps a better measure of their difficulty. In assessing a text a teacher can identify words which are likely to present difficulties to pupils, in the light of the vocabulary which teacher and pupils use in class. Some readability formulae make use of lists of familiar words, and although such lists are fairly useful, their value is limited by the fact that the passage of time affects the vocabulary in use; moreover, the best-known word lists are of American rather than British origin, and the two language forms are not identical in their vocabulary and word use.

Readability of the vocabulary must be considered in the light of the way the words are presented in the text. A word which appears in isolation (as quite often happens in mathematics) is harder to read than if it is part of a sentence, or if there are context clues or illustrations which throw light on the meaning.

Finally, the way in which the text *helps* the reader to understand its vocabulary is a very important factor. If the author takes care in providing this help, it may be possible for pupils to tolerate quite difficult vocabulary. Williams (1981) surveyed the ways in which textbooks in different content areas deal with the explanation of words. This explanation is termed *lexical familiarisation.* Many of the techniques mentioned by Williams apply to mathematics; some of the pitfalls of these techniques also apply. Williams suggests that the following checklist might be used in examining a textbook for its style of lexical familiarisation.

(a) Is it clear to the reader *when* a lexical item is being familiarized? In other words, does the book have a typographic system for familiarization?

(b) Is this system *consistent*?

(c) Does the book contain an index, and does that index distinguish (again by a typographic device) between lexical items that are familiarized in the main body of the book, and those that are not?

(d) Are the lexical familiarizations likely to do their job *for the target readership*?

(e) Are the familiarization forms chosen the most appropriate? In particular, has the author made sufficient use of non-verbal devices?

(f) If illustrative familiarizations are included, what degree of interaction between illustration and text is there, and how successful is it?

(g) Are familiarizations followed up – by presentation in a different context, in an exercise, in a diagram, in an end-of-chapter summary, etc.?

(h) If the book contains a glossary, is the language of the glosses suitable for the target readership?

(i) Are there instances of *non*-familiarization of key terminology?

(Williams, 1981)

To sum up the argument of this section: instead of using a 'long words are difficult' strategy, an assessment should be made of the difficulty of the words on a page in four ways.

1. Is the word likely to be familiar to the reader?
2. In particular, is the reader likely to have met the ME words in his mathematics lessons; do the ME words have OE meanings which might cause confusion or cause a pupil to think in terms of a limited meaning?
3. Does a particular word appear in isolation or in context?
4. Does the text have a good system of lexical familiarisation?

Assessing the difficulty of syntax

Understanding the vocabulary is only part of the difficulty of grasping the meaning of a written passage. The construction of a sentence can also pose problems to the reader. Readability formulae sometimes use *sentence length* as a measure: the longer the sentence, the harder it is to read. Again, this is only a crude mathematical model and, although it may be useful, the length of a sentence is not the only syntactic cause of reading difficulty; it may arise from the order of the words and the syntactical complexity of the sentence, and these cannot be measured by readability formulae.

Research reviewed by Perera (1980) indicates that sentences of the same length may vary greatly in difficulty and that some long sentences are easier to understand than some short ones. Evidence was also quoted to indicate that:

> . . . children read more easily those sentence structures that they would themselves say or write than sentence patterns which occur predominantly in literary writing.

The simplest structure is that of the basic *subject-verb-object* type; stylistic conventions which rearrange this structure cause problems. 'Oranges cost 4 pence' and 'He bought 2 oranges' are sentences which have a simple structure. However, the following examples show how easily this simple structure can be altered:

'Along the road came the car.' 'Listed below are the coordinates of each point.'
'At $x = 3$, the lines cross.'

Botel, Hawkins and Granowsky (1973) listed types of sentence which cause syntactical difficulty. They give a 'difficulty score' for each type, and this accumulates to provide a complexity rating for a particular passage. Such a numerical measure is no doubt beset with practical problems of use and interpretation, but the structures listed by Botel *et al.* provide useful guidelines for the assessment of a piece of writing. The simplest constructions are short sentences of the subject-verb-object or subject-verb-adverb type. Also at this level of complexity are simple rearrangements of these structures into questions or instructions; simple combinations are included in this type. Some examples are:

'Draw a circle.' 'What is the total cost?'
'The answer is π.' 'Calculate the area and find the perimeter.'

The next level of complexity is that in which simple structures such as those shown above are extended by the use of adjectives, phrases and participles. Sentences containing conjunctions other than 'and' are also at this level. Some examples are:

'Fold a piece of paper in four.'
'Coloured card or heavyweight cartridge paper are the best materials for making models.'
'If a friend walked with one of the duellers but stopped after 10 paces, where would he be on the diagram?'

Among the features which contribute to the next level of complexity are the use of the passive ('The ball was hit by Bob'), subordinate clauses ('He left *when he had finished*') and comparatives ('3 is *more than* 2'). Some examples of this level of complexity are:

> 'Carpet which costs £4.25 per square metre is to be laid in a room 10 metres long and 6 metres wide.'
> 'Use your geometry instruments to draw accurate nets of the following shapes.'
> 'Is the product of 3 and 4 more than the sum of 3 and 4?'
> 'A quadrilateral whose four corners lie on the circumference of a circle is called a cyclic quadrilateral.'

The structure which Botel *et al.* regard as the most difficult is that in which a clause is used as the subject of a sentence:

> 'The fact that the ruler is three times as long as the pencil tells us that we can write the following equation.'

Any systematic way of classifying syntactic complexity will be complicated, but the above examples illustrate the kind of sentence which pupils may find very difficult.

Research by Linville (1976) indicates that pupils find the difficulty of the syntax as important as the difficulty of the vocabulary when they are doing 'word problems'. He gave pupils problems which had easy words and easy syntax; for instance:

> 'A tree was 295 inches tall. It grew 314 inches. How tall is the tree now?'

There were also problems which had easy words and difficult syntax, such as:

> 'If you drove a car 295 miles in one day and had driven 314 miles the day before, how many miles would you have driven in both days?'

Problems containing difficult words and easy syntax, and problems containing difficult words and difficult syntax were also used. It was found that sentences with complex syntax caused reading difficulties as great as those caused by difficult words. In fact, both vocabulary and syntax were highly significant for success in solving word problems.

There are several characteristics of ME which tend to produce very complex sentences. The passive voice is commonly used, and is sometimes inescapable; 'each side of the equation is divided by 3' cannot easily be turned into the active voice since it is not clear who is doing the dividing. Some authors introduce people into the text to avoid the passive voice: 'Mary divides each side of the equation by 3', but this is rather artificial.

Mathematical authors often need to write *about* principles, equations or inequations; the statements discussed are themselves sentences:

> 'The inequalities $1 < 2$ and $2 < 3$ can be combined as $1 < 2 < 3$.'

The symbol '$<$' represents the words 'is less than'; thus, in vocalising the statement, a very complex sentence is formed. Similarly the structure of the following sentence is very complicated:

'$3x + 4 = 7$ implies $3x = 7 - 4$'

Logical implication, such as that in the above example, always causes complexity. The equivalent way of stating an implication, using 'if . . . then . . .' produces equally complex syntax:

'If tomatoes cost 53 p per kilogram then two kilograms will cost £1.06.'

The use of 'if' in a logical sense in this sentence is appropriate, but many writers have the habit of using 'if' when it is not part of a logical implication, but merely a stylistic device; the word 'we' and other 'mathematical-sounding' phrases are similarly used, as in the following example:

'If we examine the square root tables we find we have an added problem because we either have:

(a) two tables, one for the square roots of numbers from 1 to 9.9 and one from 10 to 99,

or (b) two answers printed.'

Munro (1979) discussed another aspect of mathematical syntax. The use of a ME word may affect the reader's perception of the syntax of a sentence. Munro's example is the phrase '8 divided into 2'. In ME, this may conventionally mean that 2 is divided by 8, giving an answer of $\frac{1}{4}$. The ordinary use of 'into' would suggest that 8 objects are to be divided into 2 piles, giving an answer of 4. However, if we say '4 divided into 12', the mathematical usage seems more attractive: people often say '4 divided into 12 gives 3' or '4 into 12 goes 3 times'. However, the interpretation in which 4 objects are divided into 12 piles could still reasonably be inferred, and this would give an answer of $\frac{1}{3}$. Thus, there is an ambiguity in the phrase '8 divided into 2' – and it seems that pupils' perception of what the phrase means may be coloured by their reluctance to perform an operation which has a fraction as its answer.

To sum up, the complexity of sentences is not reliably measured by their length, and an approach to measuring complexity by analysing linguistic structure is very difficult to implement. However, the importance of syntax is as great as that of vocabulary, and an informed assessment of the complexity of a piece of writing is a vital step in evaluating the readability of the text.

4 The reading of mathematical symbols

The coding of words and symbols

Like other forms of writing, mathematical text is expressed in a coded form, which the reader needs to be able to decode if he is to extract the meaning. However, there are two different coding systems used in mathematical text: that used for words, and that used for signs and symbols.

The code used in representing *words* is largely based on *sounds*; a letter or a combination of letters corresponds (roughly) to a sound. When a child has learnt to decode the letters, and has grasped the relationship between sounds and symbols used in the coding of words, he can approach a written word which he does not immediately recognise and make at least some attempt at reading it; that is, he can attempt to translate it into sounds which may turn out to have a familiar meaning. The methods used in the early teaching of reading are based on linking the written word with spoken sounds which are already familiar and meaningful to the child.

The code used in representing *signs and symbols* is essentially not based on sounds; a child has to use a different decoding method to read them. In many examples, the code is *pictorial*; for example the signs in Figure 4.1 are both fairly obvious, although the second conveys the intended meaning only when positioned

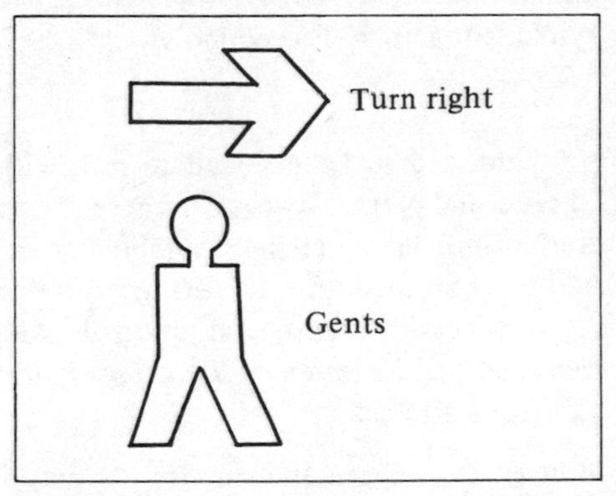

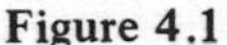
Figure 4.1

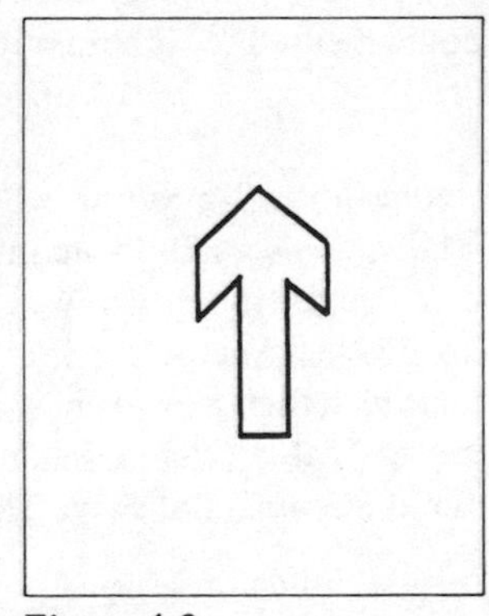

Figure 4.2

on a closed door in a public place. A further degree of abstraction is found in the direction sign of Figure 4.2, where the reader is invited to go straight ahead, and not upwards. Pictorial signs are much used in mathematics text for young children, who need to grasp the methods of coding used. For example, to interpret the illustration of the notation card shown in Figure 4.3, a child needs to realise that the figures at the bottom of the illustration are not part of the actual card, but are intended to link the concrete materials pictured above them with the number symbols whose structure is being introduced. Some primary children soon become familiar with the pictorial sign shown in Figure 4.4, which is used in the second edition of *Mathematics for Schools* (Fletcher, 1980) to tell the child to consult the teacher. However, the sign telling the teacher that a spirit master is available for use with the page is rather more *conventional* than pictorial in its coding method.

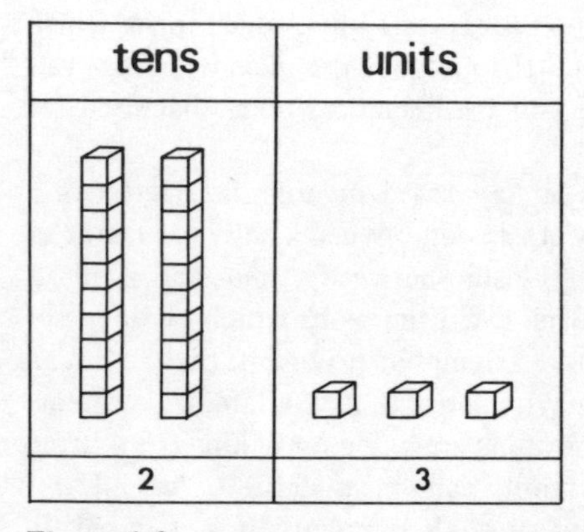

Figure 4.3

Figure 4.4

The code used in writing mathematical symbols is almost entirely conventional. There is nothing in the sound of the word 'add', or in the action of adding, which makes + a better symbol for addition than × would be, or any other mark (such as ∗) which could be used to link number symbols together. The symbols

$$3\overline{)219}$$

have a conventional meaning which is intended to be decoded to link with the action of dividing something up into three equal parts. However, because the coding method is conventional, many children do not interpret these symbols as an invitation to divide something up, but rather as an invitation to perform a numerical algorithm which they may, or may not, connect with the physical action of dividing.

Levie and Dickie (1973), in the *Handbook of Research on Teaching, Volume 2*, give a useful discussion of signs. They define a sign as:

> . . . a stimulus intentionally produced by a communicator for the purpose of making reference to some other object, event or concept.

They divide signs into iconic and digital signs. An *iconic sign* is one which in some way resembles the thing it stands for, while a *digital sign* is one which in no way resembles its referent; it is arbitrary. This use of the word 'sign' is broader than that usually used in the discussion of mathematical signs and symbols; for these authors, a word is a digital sign, and they do not use the word 'symbol'. In our discussion, we shall use the words 'sign' and 'symbol' in the rather interchangeable ways they are used in mathematics, not classifying a word as a sign, but speaking of the 'equals sign' or of 'the symbol ⊾ for a right angle'. However, Levie and Dickie's classification of signs into iconic and digital signs applies well to mathematics, although the label 'digital' is perhaps unfortunate, because digital signs do not only contain the symbols for digits. Their classification corresponds to our description of *pictorial* and *conventional* methods of coding.

Children are not always clearly aware of the difference between pictorial and conventional methods of coding. For example, the secondary school pupils discussed

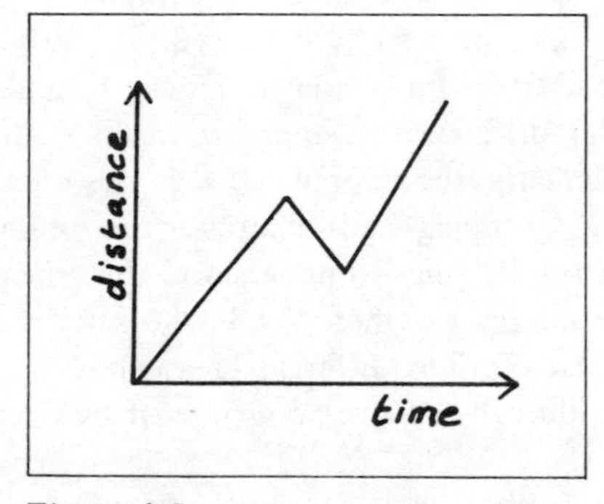

Figure 4.5

by Kerslake (in Hart 1981), who interpreted the graph of Figure 4.5 as 'climbing a mountain' or 'going up, going down, then up again', did not realise that they were dealing with mathematical symbols which use a conventional coding method, and they interpreted the graph as if it were a pictorial symbol for a mountain.

Ideas, words and symbols

There are many difficulties in interpreting the conventional coding method of mathematical symbolism; because the coding of symbolism is not based on sounds, it is necessary for children to learn the conventional meaning of each symbol or combination of symbols as they meet them. In fact, when learning symbols, children need to link together three things: an idea, some spoken (or written) words which correspond to that idea, and the symbol.

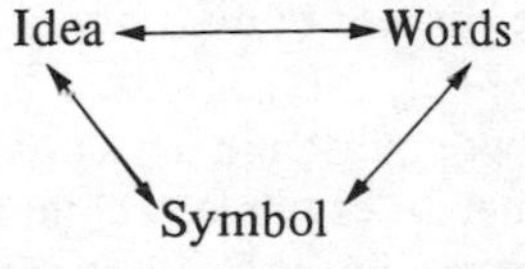

Many of the reading difficulties which children encounter arise from the complicated nature of the interrelationships between these three.

First, the correspondence between a written word and its spoken equivalent is one-to-one, while the correspondence between mathematical symbols and their spoken or written words is not one-to-one. The symbols $5 + 2 = 7$ can be read in many ways, using 'add', 'plus', 'more than', and many others. The most appropriate words depend on the context, and the context is often missing in the classroom and in the textbook, whereas in everyday speech the context is always present, and the language is appropriate to the context; for instance, the child has a 5p piece and a 2p piece, and says that he has 7p *altogether*.

The same symbols can carry more than one meaning or context. Matthews (1981) showed 11-year-olds the symbols

$$29 - 15 = 14$$

and found that under 20 per cent of them could complete a story for this operation which started: 'I picked 29 apples and you picked . . .', although 75 per cent could complete the story 'I picked 29 apples and gave away 15 and . . .'. The symbols did not carry the less familiar 'difference' meaning to the children, but only the familiar 'take away' meaning. Similarly, the symbols $10 \div 2$ are sometimes used to represent 'how many twos in 10?' (grouping) and sometimes to represent 'divide 10 into two equal parts' (sharing). In order fully to understand the concept of division, a child must link grouping and sharing together. Yet in the early stages the same symbols are apparently used for two entirely different ideas which happen to yield the same number in the results, although one gives '5 groups of two' and the other 'each part contains 5'.

In fact, each of the symbols used for the 'four rules' has multiple meanings. Children first learn how to use the symbols $3 + 4$ when they put together a set of 3 objects and a set of 4 objects. Later the same + symbol is used to describe continued measuring along a line – a length of 3 cm followed by a length of 4 cm will be described by 3 cm + 4 cm. In the case of multiplication, a child who has learnt that 3×2 means '3 sets of 2' has to extend his idea of the meaning of $\times$ considerably before he can make sense of $\frac{1}{3} \times \frac{1}{2}$.

Both the same idea and the same words can be represented by different symbols; there are three or four different symbolisations of division in use in schools:

$$6 \div 3, \quad \tfrac{6}{3}, \quad 3\overline{)6}, \quad 6/3.$$

Every pupil meets the first three of these, while the fourth is used occasionally in printed text, and always in computing. A child needs eventually to understand that there are two ideas (grouping and sharing) and several sets of symbols attached to the word 'division', including the notation $\frac{6}{3}$, which makes the link between division and fractions explicit. A child has a fully operational concept only when all these ideas are seen as aspects of the concept of division, rather than as isolated ideas and notations.

Making the connections between different systems of symbolisation is sometimes a very important step in learning; for example, a child needs to realise that the same words 'three hundredths' can be symbolised by both $\frac{3}{100}$ and 0.03, if he is to understand the links between the fractional and decimal systems of notation. At a later stage he needs to realise that the indefinite integral notation and the antidifferentiation notation express the same idea:

$\int 2x\,dx = x^2 + c$ means the same as: if $\frac{dy}{dx} = 2x$

then $y = x^2 + c$.

There is even a symbol which takes its meaning entirely from its context; this is the arrow $\longrightarrow$ representing a relation, which is now widely used in school mathematics. Sometimes words are written above the arrow to give the relation a particular meaning, as in the following example (Scottish Primary Mathematics Group, 1975):

Complete: Shape □ $\xrightarrow{\text{'has a smaller area than'}}$ shape □

In this example the child is intended to write in the boxes the names of two given shapes chosen from a set illustrated on the page, and the arrow appears unnecessary; however, later examples in the same set could be of the form

Shape □ $\longrightarrow$ shape □

because the meaning of $\longrightarrow$ has been defined as 'has a smaller area than'. This notation makes explicit to the child the fact that he is using a relation between members of the set of shapes. In other cases the child needs to know that the arrow stands for a particular relation defined in the exercise he is doing at the time, and he is expected to be able to supply its meaning in that context. The following example illustrates this (Scottish Primary Mathematics Group, 1975):

In the finished pattern,
the white part $\longrightarrow$ the shaded part
Above the arrow, write what it means.

Combinations of symbols

Another characteristic of mathematical symbols is that a symbol rarely occurs alone; symbols can be combined in various ways and it is usually the combination of symbols which conveys a meaning. The positioning and spatial arrangement of a combination of symbols is all-important in conveying a meaning, and the meaning of a combination of symbols is governed by convention in just the same way as the meaning of a single symbol is conventional; 41 and 14 stand for different numbers, and at some time in history an arbitrary decision was made to write the tens on the left of the units. This may or may not fit the conventions of speech in a particular language; the left-to-right order of the symbols in 41 matches the order of the words 'forty one', but the symbols 14 do not match the order of the components of the word 'fourteen'.

There are very many other irregularities in the rules for combining symbols, and children cannot work these irregularities out for themselves, but have to learn the conventions. The symbol for 'centimetre' is cm, and the symbol for 'three centimetres' is 3 cm; the symbol for 'a quarter' is $\frac{1}{4}$, but the symbol for 'three quarters', which has exactly the same structure, is not corresponding graphic form $3\frac{1}{4}$, but $\frac{3}{4}$. The symbol $3\frac{1}{4}$ symbolises the idea 'three and a quarter'; the principle of combining symbols in this case is addition rather than the principle of naming the unit which is used. Further confusion may arise from the fact that the words 'three and a quarter' must sound to many young children very like the words 'three quarters', though the two sets of words convey different ideas.

Later, when he starts to learn algebra, a pupil learns that the method of coding used in the combination of symbols $3x$ is not same as the coding used in the combination of symbols 34. However, in $3x$, the value of x may sometimes be 4, but the principle of combining symbols is multiplication, rather than the 'tens and units' coding used in 34. A pupil who wants to give $x = 9$ as the solution to $3x + 1 = 40$ is making the wrong choice between these coding systems; he may also complain that $3x + 1 = 25$ 'doesn't make sense'.

When a pupil has got to grips with the fact that $3x$ stands for '3 multiplied by x', he is starting on a road which eventually leads to calculus, where he will learn that δx is not a symbol for δ multiplied by x, but for 'an increment in x' and, in spite of the superficial resemblance of $\frac{dy}{dx}$ to $\frac{ay}{ax}$, the a's may be cancelled out in the second set of symbols, but in no circumstances may the d's be cancelled in the first set. The reading of mathematical symbols might be likened to reading a language in which all the verbs are irregular and all their forms have to be learnt separately. However, difficult though the process of learning to read and write mathematical symbolism is for many children, it is a most important part of learning mathematics, and a child will not make good progress in mathematics unless he builds up his knowledge of its symbolism alongside his understanding of its ideas. In this context, we recall that Kane, Byrne and Hater (1974) found that the only mathematics tokens which were known to 90 per cent of American children in grades 7 and 8 (ages 13 and 14) were +, −, ×, \$, ÷, %, ¢ and the numerals.

The work of the APU (1981) has shown that large numbers of pupils fail to understand the coding system used in algebra. Only about 45 per cent of 15/16-year-olds were successful in a pencil-and-paper test on each of the following questions:

> n represents a whole number.
> How do we represent the number which is:
> one bigger than n?
> three less than n?
> twice n?

In answer to the last question, 21 per cent of pupils gave n^2 as their reply. In answer to the question:

'$y = d^3$. Find y if $d = 3$.'

the answer 9 was given by 19 per cent of pupils. These pupils seem to think that the index notation stands for the multiplication of the letter by the index, so that $n^3 = n \times 3$. They may also think that $3n$ stands for $3 + n$, because when in the practical testing they were asked to estimate v, given that $v = u + gt$, and given values for u, g and t, about one-third of the pupils added the three numbers. It would seem likely that considerable numbers of pupils who are among the candidates for CSE or even O-level have not understood the method of coding used in algebraic symbolism.

It is not always clear where to start to read a combination of symbols; we read prose text from left to right along each line, and from the top to the bottom of the page. This convention does not always apply to the reading of mathematical symbols. The long division algorithm is a particularly difficult example of this, as the following example of a child's attempt to reproduce an algorithm shows. Luther, aged 8, was observed trying to do what he thought was required when working on a page of his textbook which included exercises on division. He mistakenly started the process by writing down what he thought was the answer:

```
  44 r.2
2)99
```

and then completed the sum with the 'decoration':

```
  44 r.2
2)99
  90
   9
   9
   0
```

On questioning, he explained what he had done as:

2 into 9 goes 4, remainder 1,
2 into 9 goes 4, remainder 1,
so the remainder is 2 altogether.

This made it clear not only that Luther did not understand what the long division process meant, but also that he thought he should fill in his working of the process by starting at the top line and working downwards to achieve a 'sum' of the right shape. The written long division process will not make sense to him until he not only understands what dividing means in a physical sense, but also sees the relationship between the physical process of dividing and the complex pattern in which the written procedure is constructed.

The order and arrangement of symbols is important in many other places in mathematical writing as well as in the place-value system for writing numbers. For example, both the combinations of symbols $6 \div 3$ and $3\overline{)6}$ convey '6 divided by 3', but the ordering of the symbols is opposite. It is not surprising that some children make the generalisation that in all division, the larger number must be divided by the smaller. Thus, CSMS (Hart, 1981) found that in response to the question 'Divide by twenty the number 16', 23 per cent of fifteen-year-olds chose the reply 'there is no answer', and one of the pupils interviewed simply replied to this question, 'You can't.' In the case of another symbolisation of '6 divided by 3', $\frac{6}{3}$, a child was found to write:

$$\frac{6}{3} = 3$$

thinking that this notation was merely another arrangement of

$$\begin{array}{r} 6 \\ -3 \\ \hline \end{array}$$

which he had been doing on the previous page of his textbook.

Intuitive ideas of symbolism

Confusion may well stem from the fact that when children first learn to symbolise combinations of numbers, they are dealing with addition. Here the order of the symbols seems irrelevant; the child puts out a set of 3 blocks and a set of 2 blocks and pushes them together; he says 'three add two makes five' and writes '$3 + 2 = 5$'; the position of the blocks does not enter into what he is doing. In subtraction he merely takes away some of a set of blocks which he has already laid out; he may well not realise that in recording what he has done, $5 - 2$ is acceptable, and $2 - 5$ is not. The reason that he is not allowed to write $2 - 5$, when he is allowed to write $2 + 5$ and $5 - 2$, is beyond his experience at the time.

A related problem is that a symbol may not always mean the same thing to a child as it does to an adult. Some children think that 12 should be written in the box to complete the sentence:

$$3 + \square = 9$$

Research on children's understanding of the equals sign has been surveyed by Kieran (1981), and the following points are based on her work. Children's practical experience of putting sets together encourages them to think of $3 + 5 = 8$ as '3 and 5 *make* 8'; they quite rightly (at first) regard the = sign as a signal to carry out the physical action of combining the sets, or its internalised equivalent. Thus,

$$3 + \square = 9$$

does not make sense, and children think that it must have been written in the wrong order. They think that what must have been intended was

$$3 + 9 = \square$$

but that it has been written in a funny way, so they are sure that 12 is the correct number to put in the box. This problem seems to be present even among some children who verbalise the sentence as '3 plus some number equals 9'. To an adult, the symbol = is not an instruction to 'do something', but an indication that the two sides of the equation are equivalent. Thus the teacher may be talking at cross purposes with a child who understands something different by the symbol.

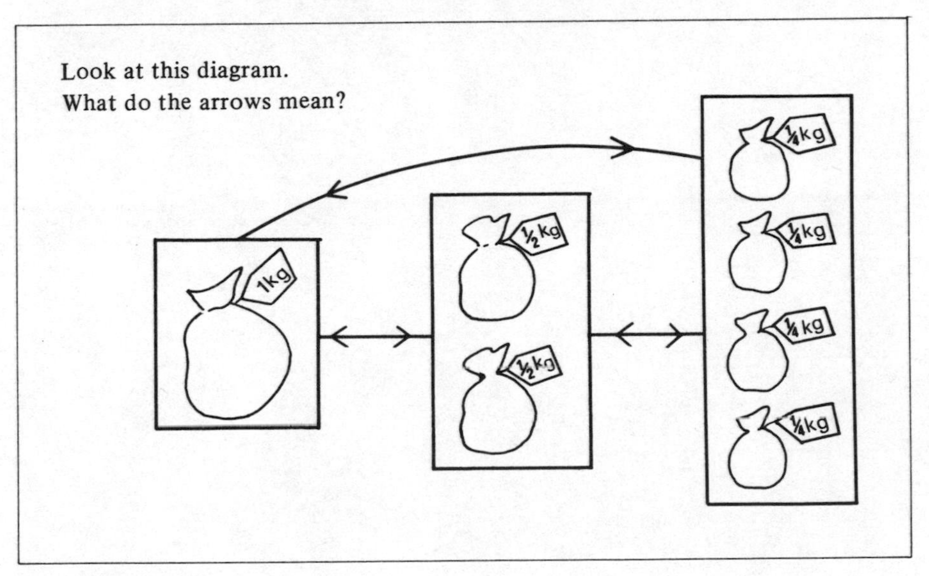

Figure 4.6

A similar 'signal to do something' interpretation of ⟶ was found in an eight-year-old's response to the question in Figure 4.6 (Scottish Primary Mathematics Group, 1975): He wrote

> 'There emptying the pan' (sic)

Another child in the same class had completely failed to grasp that the arrow was intended to signify a relationship:

> 'It means that one is poynting to the 1 kg and the other is poynting to the little kg'.

Hence it seems likely that many children will only have a partial understanding of the abstract ideas conveyed by symbols such as = and ⟶ until some years after they have begun to use these symbols. Moreover, the teacher may need to pay specific attention to the coding systems used in algebra, and in other areas of symbolism, if pupils are to interpret mathematical symbols correctly. There is no doubt that the teacher has a very important part to play in enabling children to understand the coding system of mathematics; without his help, pupils will not be able correctly to interpret text which contains mathematical symbolism.

5 Graphic language in mathematics

The visual language of mathematics

The marks on paper used to communicate ideas are appropriately called 'visual language', in contrast to the 'oral language' of speech. *Prose* forms only a part of the visual language used in the communication of mathematics; other forms of visual language used include the *mathematical symbols* which were discussed in the last chapter, and various forms of *graphic language*, such as:

- tables
- graphs
- diagrams
- plans and maps
- pictorial illustrations

The reading of these types of graphic language forms the subject matter of this chapter.

In learning to read, children are systematically taught the skills of decoding messages written in prose; however, equal care is not usually devoted to the systematic development of the skills needed for decoding messages written in symbols and in graphic language. In the communication of mathematics, the types of graphic language listed above are almost as important as prose and symbols; hence children require specific reading skills for mathematics, and the skills needed for reading graphic language are additional to those needed for the reading of prose and mathematical symbols.

Twyman's classification of visual language

The marks on paper which form the visual language of mathematics can be classified in several ways, each of which throws some light on the problems of reading mathematics. One classification is derived from the fact that a characteristic feature of prose, not shared with other forms of visual language, is its regular arrangement; in reading a passage of prose, the eye moves from left to right along the lines, and from top to bottom down the page. Twyman (1979) suggested a two-dimensional classification of visual language which takes account of the degree of regularity of eye movement needed. One dimension of the matrix shown in Figure 5.1 shows whether the arrangement is nearly linear or very non-linear, while the other dimen-

	Method of Configuration	Pure linear	Linear interrupted	List	Linear branching	Matrix	Non-linear: directed viewing	Non-linear: most options open
Mode of Symbolisation	Verbal/ numerical							
	Pictorial and verbal/ numerical							
	Pictorial							
	Schematic							

Figure 5.1 Twyman's classification of visual language

	Method of Configuration	Pure linear	Linear interrupted	List	Linear branching	Matrix	Non-linear: directed viewing	Non-linear: most options open
Mode of Symbolisation	Verbal/ numerical		Prose	Displayed formulae	Tree diagram	Table Matrix	Flow chart	
	Pictorial and verbal/ numerical		Diagrammatic instructions (strip)	Key to chart		Isotype graph	Pie chart	Venn diagram
	Pictorial						Illustration	Geometrical figure
	Schematic	Number line				Cartesian graph	Network diagram	Plan Map

Figure 5.2 Mathematical examples of Twyman's classification

sion shows the characteristic mode of symbolisation. In Figure 5.2, we have placed some of the forms of visual language found in mathematics text written for children in those cells of Twyman's matrix which seem to be appropriate. Twyman's classification scheme was designed to accommodate all possible types of visual language, not only that used in mathematics; however, it is useful for mathematics, because it draws attention to the variety of types of reading which a child needs to use in understanding mathematical writing.

It is clear that most visual language in mathematics is neither linear nor even arranged in a linear order with breaks at the ends of lines (a configuration which Twyman calls 'linear interrupted'). Hence, in reading most types of visual language the reader does not scan the page in the way that he did when he 'learnt to read', that is to say, by scanning the page from top to bottom and from left to right along each line. This method of processing is only used for linear interrupted text such as prose writing and strip cartoons. Illustrations and diagrams usually need less structured processing than does prose; the order in which the parts of a diagram are examined may make little difference to the message it conveys, provided that the whole of the diagram is read. Thus, most diagrams are non-linear in configuration. However, Twyman thinks that graphs are based on the structure of the two axes, and so he regards them as two-dimensional, and classifies them as matrices which use a schematic form of symbolisation. A flow chart, however, is an example of a diagram which needs to be read in a predetermined order, but the order is not linear.

The relation between illustrations and text

The main structure of any piece of mathematical writing is provided by the prose text; illustrative matter such as pictures, graphs and diagrams, is normally structured in relation to the prose reading matter. Illustrative matter can be classified in terms of how importantly it is related to the prose text. Three levels of importance can be found in the illustrations of children's mathematical texts:

D decorative
R related but non-essential
E essential

Decoration lightens or breaks up the text, sets the scene or fills in blank spaces on the page. It serves no instructional purpose, but is intended to make the work more attractive to read. An illustration may be entirely decorative, or the decoration may form only part of a more important illustration. In the example shown in Figure 5.3, p.48, the detailed drawing of the sailors forms a decorative adjunct to the essential bearings shown in the diagram. Nevertheless, these details perform the important function of suggesting that sailors actually use bearings in their work.

Related but non-essential illustrations can repeat and emphasise ideas stated in the prose text. Illustrations of this type may increase the number of ways in which

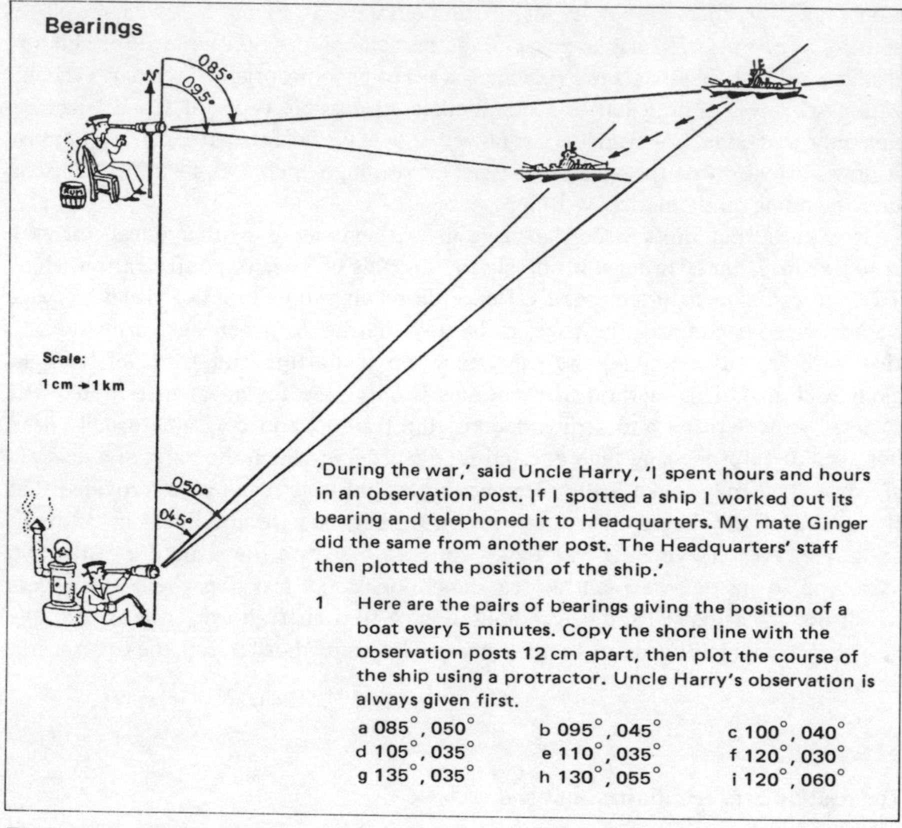

Bearings

N
085°
095°
Scale:
1 cm → 1 km
050°
045°

'During the war,' said Uncle Harry, 'I spent hours and hours in an observation post. If I spotted a ship I worked out its bearing and telephoned it to Headquarters. My mate Ginger did the same from another post. The Headquarters' staff then plotted the position of the ship.'

1 Here are the pairs of bearings giving the position of a boat every 5 minutes. Copy the shore line with the observation posts 12 cm apart, then plot the course of the ship using a protractor. Uncle Harry's observation is always given first.

a 085°, 050°	b 095°, 045°	c 100°, 040°
d 105°, 035°	e 110°, 035°	f 120°, 030°
g 135°, 035°	h 130°, 055°	i 120°, 060°

Figure 5.3 A mixture of 'essential' and 'decorative' illustration, taken from *Maths Adventure*, Book 4 (1977), and reduced to 65%.

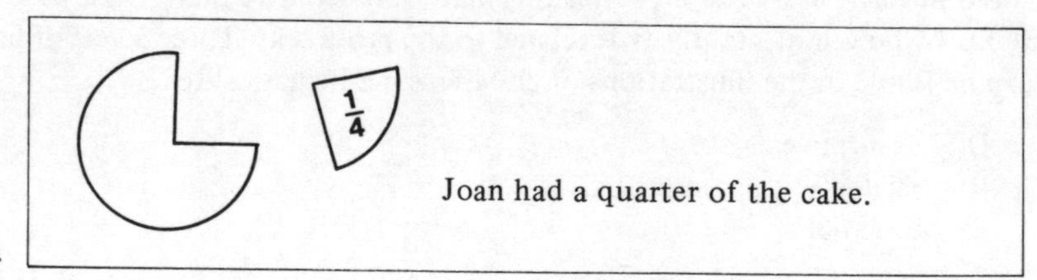

Figure 5.4

the reader can extract meaning from the text. The illustration to 'Joan had a quarter of the cake' in Figure 5.4 serves as a reinforcement of the meaning of a quarter. Price lists, such as that in Figure 5.5, are often found in primary mathematics texts; they not only break up the page, but also assist the reading of words used in the text. It is sometimes argued that the over-provision of visual support for the text causes pupils to depend too much on this support, so that they do not learn to cope with the text itself; on the other hand, clues to the meaning of the text may help the pupil to understand it, and visual material can help him with difficult words and can put difficult ideas in a concrete setting. Moreover, related

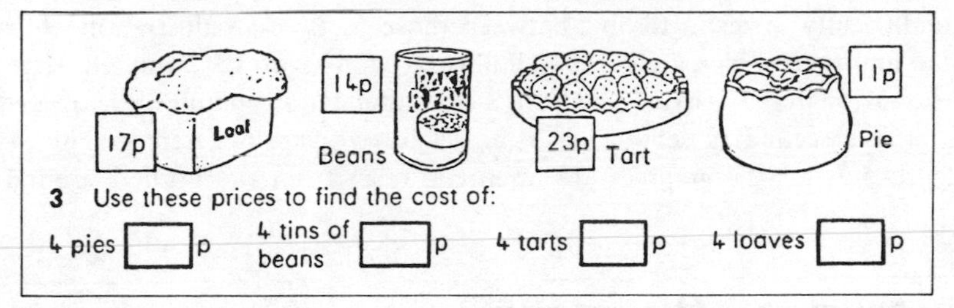

Figure 5.5 A visual price list, taken from *Primary Mathematics*, Stage 2, Workbook 2 (1975).

illustrations have an important role in giving visual embodiments and clues for abstract mathematics, so enabling the pupil to think out a problem which he might not have been able to tackle in the abstract.

Essential visual material often consists of diagrams, graphs or tables. It is not repeated in the text – indeed, often the ideas it contains cannot be expressed conveniently in words. The pupil needs to 'read' this visual material with close attention, along with the rest of the text, as it is an intrinsic part of the material provided for learning or problem solving.

The picture problem of Figure 5.6 is conveyed through *essential* visual material. An equivalent word problem:

> A rectangular tray holds 6 rows of pies, with 6 pies in each row. How many pies does the whole tray hold?

is much less clear, and demands entirely different reading skills from the picture problem.

How many pies in the tray?

Figure 5.6 Essential visual material in a picture problem (M. Holt and A. Rothery, 1981).

It may not always be apparent to the pupil which parts of the illustrative material are essential, especially if the page also contains decorative or related illustrations. In the illustration of *Bearings* shown in Figure 5.3, the decorative ships hide essential parts of the diagram, the points where the bearing lines cross; sometimes, as in this case, the designer has to make a difficult choice of whether to set the scene vividly in the illustration or to bring out the mathematics as clearly as possible.

The difficulty of distinguishing between those parts of an illustration which are essential and those which are not is well illustrated in Figures 5.7 and 5.8. Here, the same textual material, part of a draft of a text intended for pupils of below average ability in the secondary school, has been set in two very different graphic styles. In Figure 5.7, all the graphics are essential, apart from the borders around the

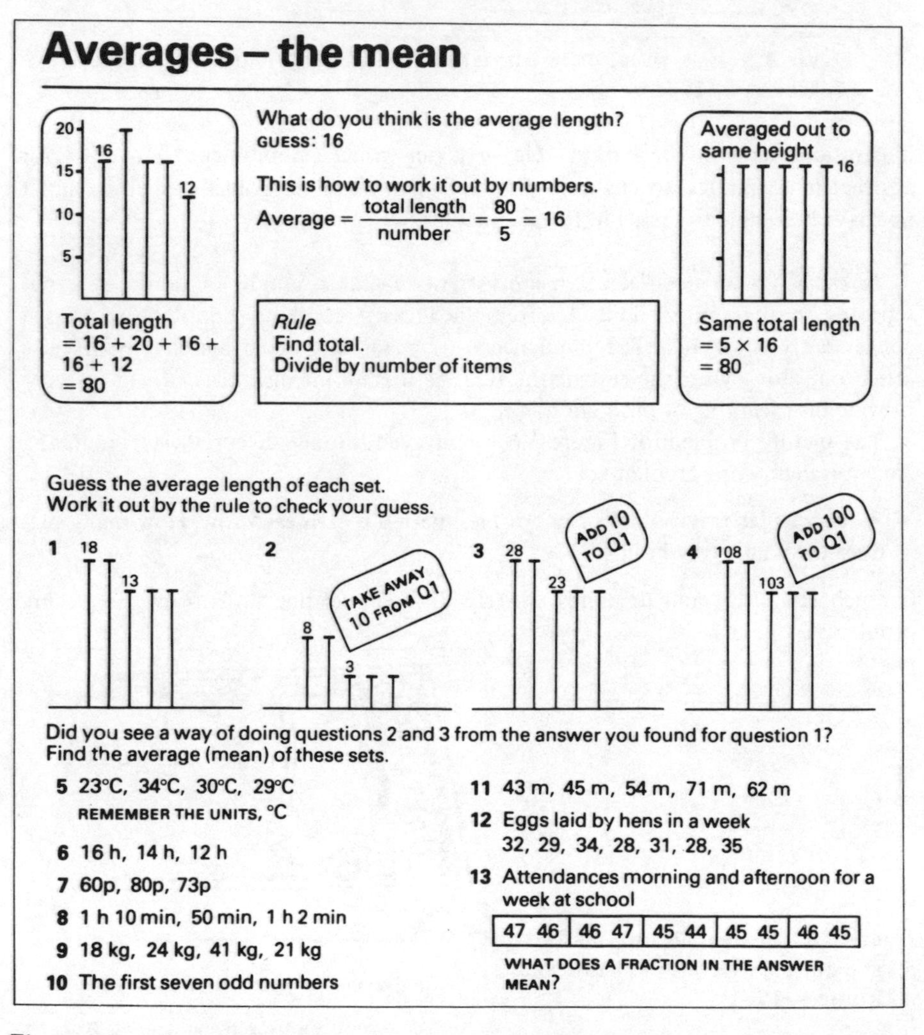

Figure 5.7 Essential material only shown in the illustrations.

boxes, which serve an important purpose in steering the reader round a closely packed page. In Figure 5.8, the essential heights of the buildings almost disappear beneath the unrelated decorative graphics. In this diagram, the three types of graphics are identified by the code letters D, R and E. In considering these two pages, the

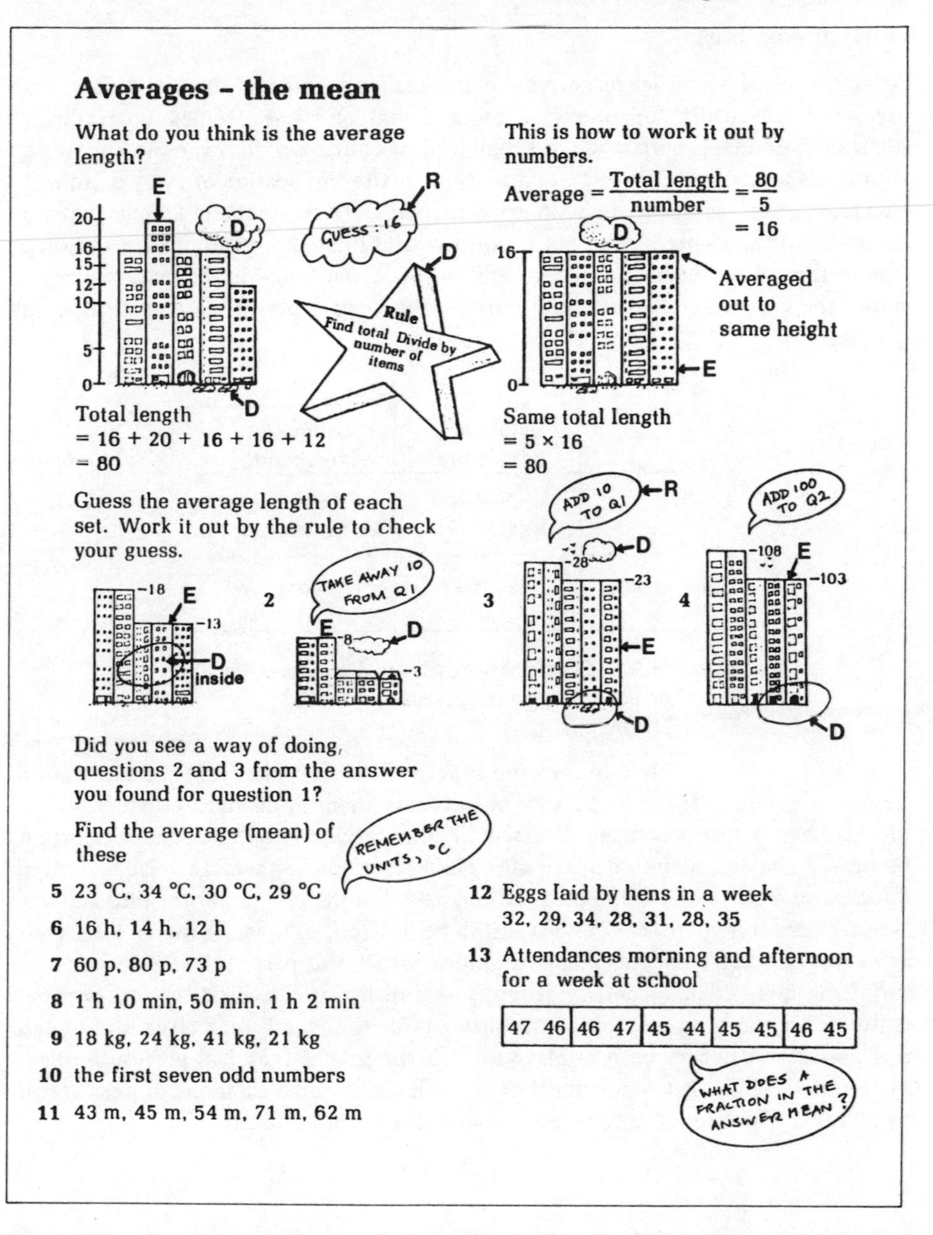

Averages – the mean

What do you think is the average length?

Total length
= 16 + 20 + 16 + 16 + 12
= 80

This is how to work it out by numbers.

$$\text{Average} = \frac{\text{Total length}}{\text{number}} = \frac{80}{5} = 16$$

Averaged out to same height

Same total length
= 5 × 16
= 80

Rule
Find total Divide by number of items

Guess the average length of each set. Work it out by the rule to check your guess.

Did you see a way of doing, questions 2 and 3 from the answer you found for question 1?

Find the average (mean) of these

5 23 °C, 34 °C, 30 °C, 29 °C

6 16 h, 14 h, 12 h

7 60 p, 80 p, 73 p

8 1 h 10 min, 50 min, 1 h 2 min

9 18 kg, 24 kg, 41 kg, 21 kg

10 the first seven odd numbers

11 43 m, 45 m, 54 m, 71 m, 62 m

12 Eggs laid by hens in a week
32, 29, 34, 28, 31, 28, 35

13 Attendances morning and afternoon for a week at school

47	46	46	47	45	44	45	45	46	45

Figure 5.8 Decorative and related material added to the illustration.

reader should take little notice of the actual text; the examples were devised by a design studio purely for the purpose of showing different styles of graphic design. Most teachers and pupils seem to prefer the first presentation; what it lacks in motivational appeal it gains in visual clarity.

Levels of processing

Some pieces of visual language require the reader to process them in careful and precise detail, while for others, a more global and less detailed processing is sufficient. An inappropriate method may lead to confusion; for example, the reader who reads a light novel with grave attention to the implication of every word may well lose the thread of the story. In mathematics, both the text and graphic material need very detailed processing, but decorative and related illustrations are an exception to this, as are the 'metalanguage' and signals often found in the text. Figure 5.9 shows the processing methods appropriate to various types of mathematical visual language.

	Detailed processing	Global processing
Text	Exposition Exercises	Metal language Signals
Illustrations	Essential	Decorative Related

Figure 5.9 Classification of mathematical visual language according to processing method

It is difficult for a child to distinguish between the types of text and illustration shown in Figure 5.9, and so he may not process them in the most efficient way in his reading. If the processing of essential illustrations is not detailed, important points may be missed; in order to ensure close attention to them, the reader is often directed to a particular detail by questions asked in the text about the illustrations. Nevertheless, essential features may still be missed, as was shown in Kerslake's survey of children's understanding of graphs, which was part of the CSMS study. A surprising number of secondary school children had not realised that an essential feature of a graph is a constant scale throughout the length of each axis, although this feature must have been displayed on all the graphs they had previously read. Figure 5.10, which is taken from Hart (1981), shows some examples of axes drawn by children who had failed to realise the importance of this feature.

Eye-movements

The characteristic eye-movement in reading ordinary prose, is from left to right along each line, and from top to bottom of the page. Illustrations break up this regular eye-movement, and essential illustrations may not be properly studied unless they are comfortably positioned in relation to the text references. There seems to have been little research into the positioning of illustrations in mathematical text, but Whalley and Fleming (1977) have experimented with rearranging the

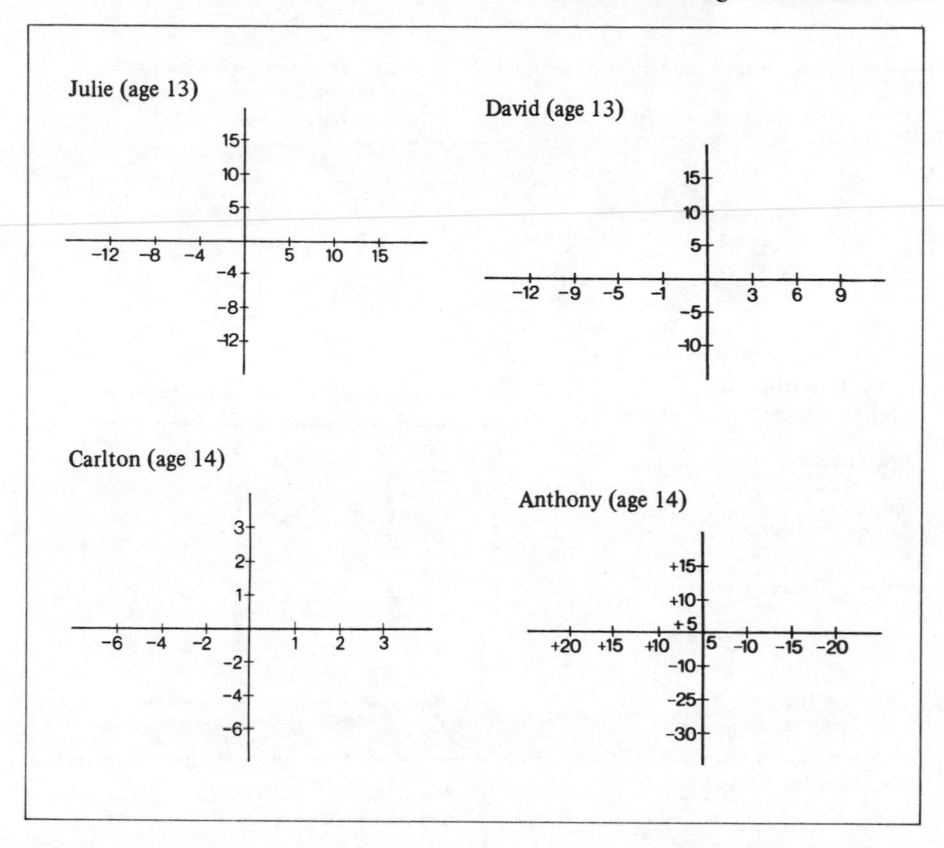

Figure 5.10 Examples of children's inappropriate labelling of the axes of a graph.

diagrams in an illustrated technical article on electronics: they rearranged the text so that each diagram immediately followed the line of text in which it was first referenced. In the original version, diagrams were arranged to 'balance' the text. Whalley and Fleming compared the effect of the original and the rearranged text; the subjects were students of electronics, and an electrical device was used to record the places in the text on which the subject focussed during reading. The subjects who worked with the rearranged text spent 35 per cent of their reading time on the diagrams, as against 15 per cent for those who used the original text; none of the subjects looked at two diagrams which were on the 'wrong' page of the original text. The subjects who used the rearranged text also reported significantly more often that the text was clear, and that they understood it. It is not surprising that students who spent more time reading diagrams should feel that they understood the text better, especially if the diagrams were originally conceived as an essential part of the learning material. The rearranged text made it as easy as possible for the eye to move backwards and forwards between the text and the diagram, and to ensure that the diagram was in a predictable place in relation to the text.

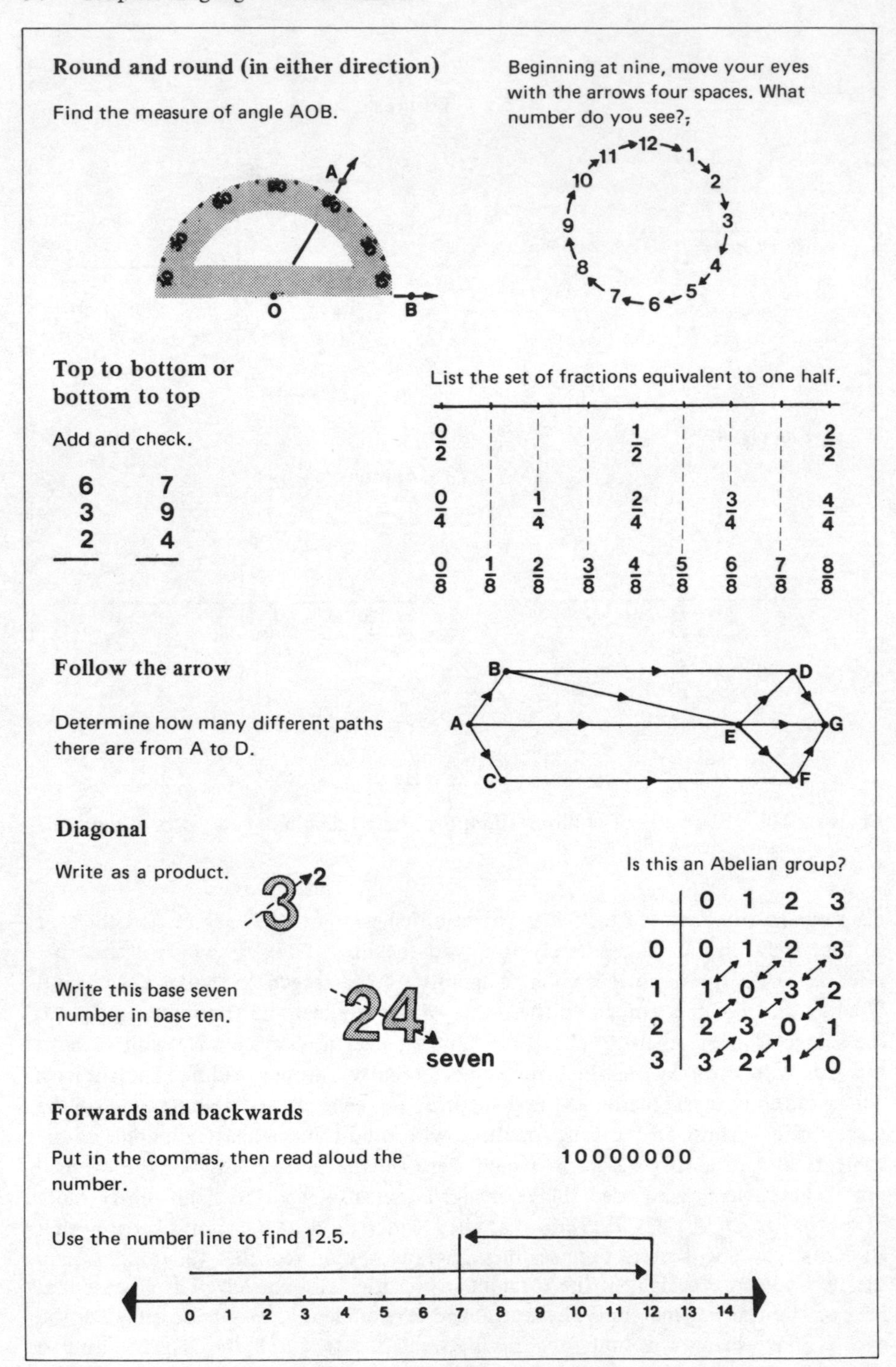

Figure 5.11 Eye-movements used in reading a mathematical diagram.

The variety of mathematical illustrations is such that there is no universal route which eye-movements could take, which would serve for the reading of every illustration. The Office of Instructional Services of the Georgia Department of Education has published a very useful booklet, *Reading Mathematics* (undated); a large part of this consists of mathematical illustrations that require different eye-movements in reading. The types of eye-movement identified in this booklet are shown in Figure 5.11.

In these examples it is worth noting that in each case the diagram is an essential part of the text and that no part of the diagram is decorative or related; yet the pupil often only needs to read part of the diagram in order to answer the associated question. For example, a pupil who correctly lists 'the set of fractions equivalent to one-half' may not have noticed the other equivalences shown in the diagram.

Material written for young children often requires them to undertake a third task alongside the reading of text and illustrations; the text frequently asks the child to carry out some practical task, part of which is illustrated on the page. In

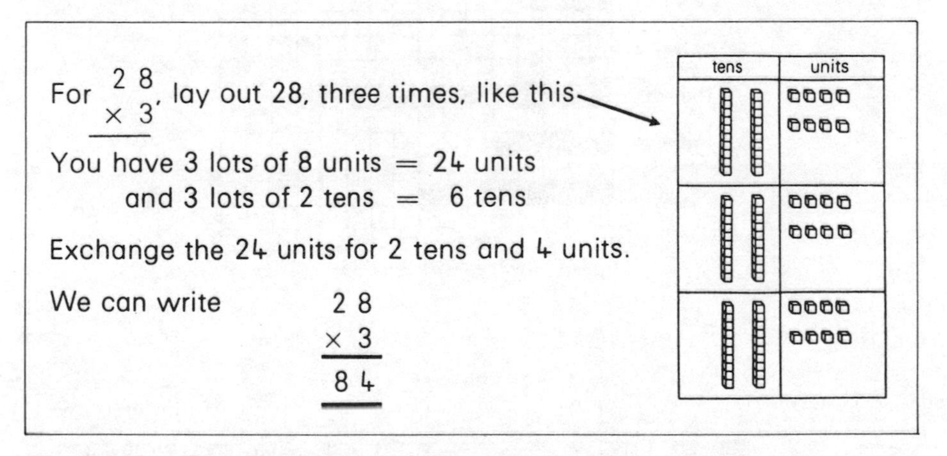

Figure 5.12 Text, illustrations and apparatus all needed.

the extract shown in Figure 5.12, taken from *Scottish Primary Mathematics Group,* Stage 2, Workbook 2 (1975), the original arrangement of the number apparatus is illustrated, but not the later stages in its manipulation; the text comments on later stages in the manipulation of the apparatus, which it assumes the child is carrying out correctly.

Conventions of shading, colouring and scale

In a chapter entitled 'The graphic language of organisation', Meredith (1961) has called attention to the variety of codes which are used in organising ideas graphically. His book is a study guide for adults who are starting a course of independent study, and he describes for these students the types of graphic language used in organising and summarising ideas; his views are also very helpful when examining the illustrations in mathematics texts.

> One of the factors which makes graphic organisation so powerful is that it can draw simultaneously on a number of different codes and so achieve great economy of expression. . . . Some codes are more obvious than others. Some are so little obvious as to be invisible. The most important example of this tacit symbolism is the relative position of the elements in a diagram. The centre, the top, the bottom, the sides, the proximity or distance of the elements, all these can and should signify important relations.
>
> (Meredith, 1961)

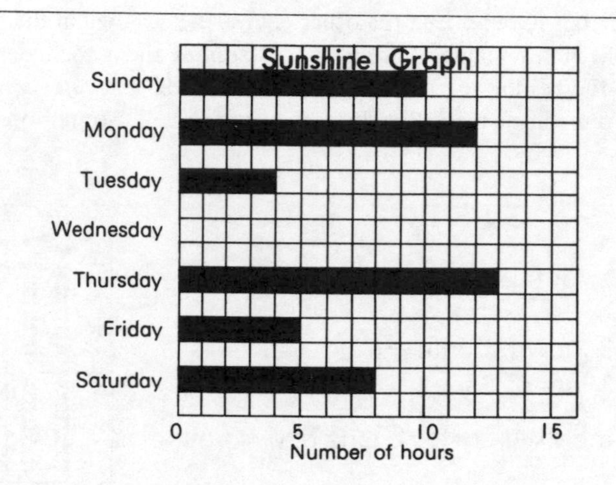

Here is a graph showing part of the 3 times table:

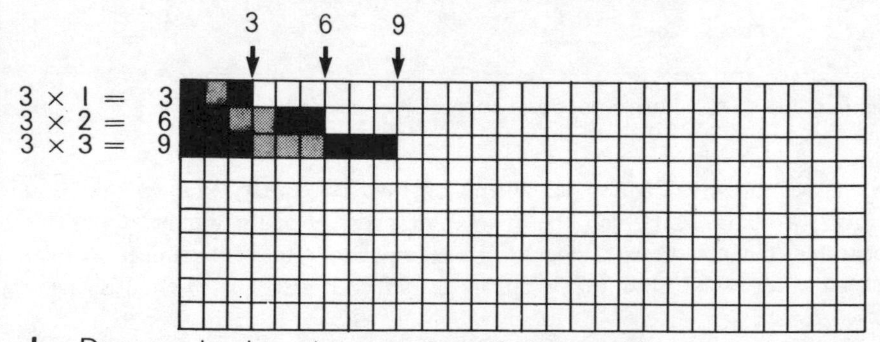

1 Draw and colour the rest of the 3 times table.
The numbers 3, 6, 9 along the top are the first three answers.
Write in the other answers.
Complete the 3 times table at the left of the graph.

Figure 5.13 Different uses of colour in graphs, indicated here by variable shading.

Though the codes used for presenting and organising ideas in a mathematical diagram are familiar to the teacher through long experience, these codes may be less obvious, or even undetectable, to the pupil. There seems to be no recent research on children's understanding of diagrammatic matter; however, Malter (1948) found that many elementary school children, at stages between grade 4 and grade 8, did not understand the significance either of using broken lines in a diagram to indicate a position from which a moving part had come, or of using a serrated line to indicate that a part had been cut away so that the interior could be shown. Children's books have changed in the last thirty years, as has the provision of illustrated material in books and on television; however, Malter showed that it cannot be taken for granted that children will automatically infer for themselves the significance of diagrammatic conventions.

Every mathematics text contains diagrammatic conventions, examples of which can be found on almost every page. Shading and colouring, for example, play a number of different roles; these often vary between diagrams in the same text. In *Primary Mathematics,* Stage 2, Workbook 2 (1975), the graphs shown in Figure 5.13 appear on consecutive pages. In the first graph the bars are coloured green. This colouring merely picks out the bars in contrast to the background, whereas in the second graph, two shades of green are used to indicate ones, twos, threes, and so on, in building up the multiplication table. This convention is easily missed, and is not commented on in the text.

In work on fractions, shading and colouring are almost essential, and diagrams such as those shown in Figure 5.14, taken from *SMP Book D* (1978), are typical.

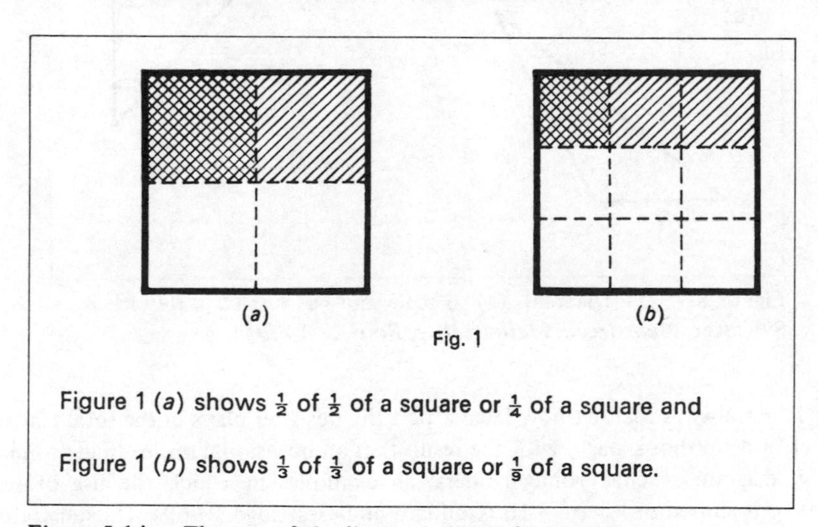

Figure 5.14 The use of shading to indicate fractions.

To the child who knows the convention, Figure 1(a) in the extract certainly shows $\frac{1}{2}$ of $\frac{1}{2}$ a square, but another child may only see a square divided into four smaller squares, one of which is shaded in red (in one direction), while another is shaded in red and black (in both directions).

Scale plays an important part in many geometrical diagrams. Dimensions are often marked on diagrams, and a diagram sometimes has the same dimensions as those labelled, or is carefully drawn to some convenient scale. However, this is not always the case, so that a pupil who tries to make his own drawing will obtain a diagram of a different shape from that in the text. In the two diagrams shown in Figure 5.15, from *Mainstream Mathematics*, the first is drawn accurately to scale, while the triangle in the second diagram is not of the correct shape. As they progress through the secondary school, pupils are expected to learn to distinguish diagrams in which the scale is important from those in which a sketch (not to scale) is used.

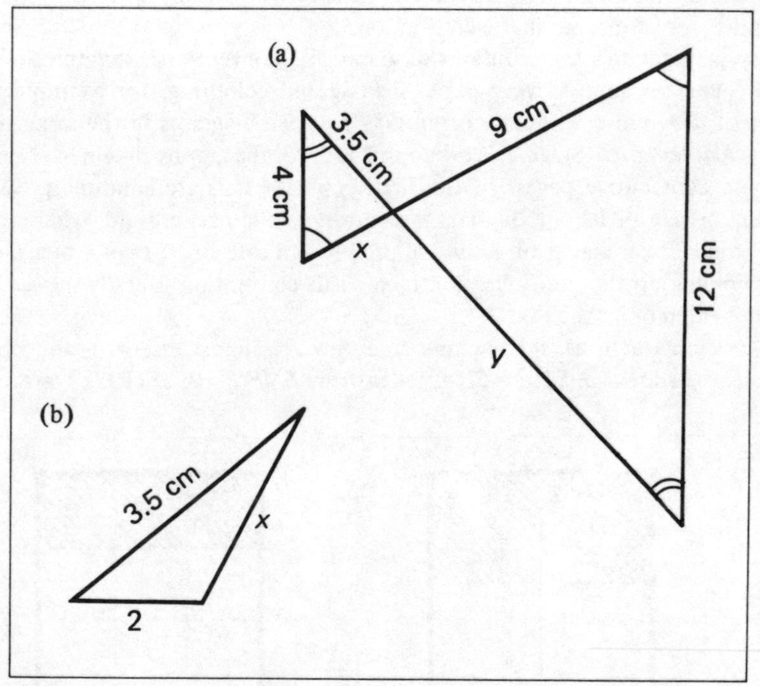

Figure 5.15 Diagrams (a) to scale and (b) not to scale (J.E.K. Sylvester, *Mainstream Mathematics, Book 2,* 1979).

It is not always realised how large a part the designer plays in the total teaching impact of a textbook page, with the result that an occasional misleading, or indeed wrong, diagram reaches young readers; an example, in which the use of scale misleads, is shown in Figure 5.16 (Goddard and Grattidge, 1969). The illustrator's portrayal of the litre and half-litre measurements at the top of the page is a notable example of the misrepresentation which can be caused by confusing one-dimensional measurements with two- or three-dimensional measurements in the conventional symbolisation of quantities. Moreoever, each of the other objects on the page is drawn to a different scale.

Figure 5.16 Misleading use of scale.

Sometimes a diagram is used to make a mathematical point in a semi-realistic pictorial style such as that shown in Figure 5.17 (*Mathematics for Schools,* Level II, Book 1, second edition, 1980). In order to show that 2 is the input to the function machine and 8 is the output, both number cards need to be shown, but the diagram then has to break the convention that a realistic picture portrays a single moment in time; 8 could not emerge from the machine before 2 went in.

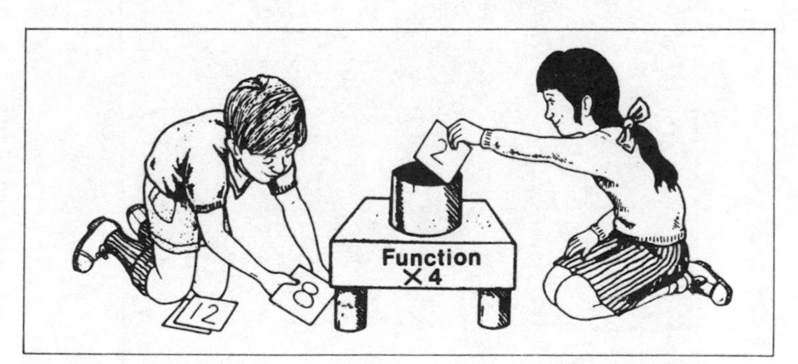

Figure 5.17 Breaking conventions of time in an illustration.

Conventions of motion

In pictorial illustrations used in mathematics text for young children, it is sometimes necessary to indicate that objects in the picture are in motion. For instance, the picture may show that one set of objects is to be combined with another, or it may help the young reader to grasp the meaning of words such as 'round and round' which occur in the text. Campbell (1981) has studied children in the first grade in the USA (aged about 6), to see how they interpret pictures which use conventional indications of motion to show the joining together or partitioning of sets. An example of the type of picture she used is shown in Figure 5.18. Children were asked to tell stories about the pictures; not all the children saw the mathematical significance of the pictures when they told stories about them. One child responded to the pictures of ducks with 'Two ducks. Two ducks. Four ducks. Six ducks.'

Figure 5.18 Use of motion to show addition.

Another child told the following story about a picture which showed two crabs running away from a group of four crabs:

> 'Some guy's trapped on an island and he wants some food, so he's trying to catch some crabs to eat. He can't get off the island 'cause there's some sharks swimming around the island.'

Neither the man nor the sharks were shown in the picture.

Campbell was able to distinguish four stages in the interpretation of her pictures:

- a general story was told about some object in the picture
- the objects were grouped in sets, but there was no recognition of motion or of the mathematical relationship between the sets
- motion was recognised, but the mathematical relationship between the sets was not
- the picture conveyed the mathematical relationship intended

More generally, the psychological research discussed by Campbell has been able to distinguish three stages in children's interpretation of pictures:

(i) the child keys in on some single item of the picture
(ii) the child lists the objects in the picture
(iii) the child begins to see motion and relationship in the picture

Two different conventions are used to show that figures or objects in a picture are in motion. Sometimes *postural* cues are used, so that figures are shown off balance or in a walking posture. The other type of cue is *conventional*: lines or clouds of dust are shown following the moving figure. The middle picture of Figure 5.18 shows both types of cue.

British texts use the types of pictures studied by Campbell less often than the more lavishly illustrated American books, but some similar examples are to be found. For example, the first two pictures shown in Figure 5.19, taken from *Mathematics for Schools*, Level II, Book 1 (1971), use motion to indicate 'taking away'. In the next two pictures, symbols of carrot-tops and acorn cups are used to indicate

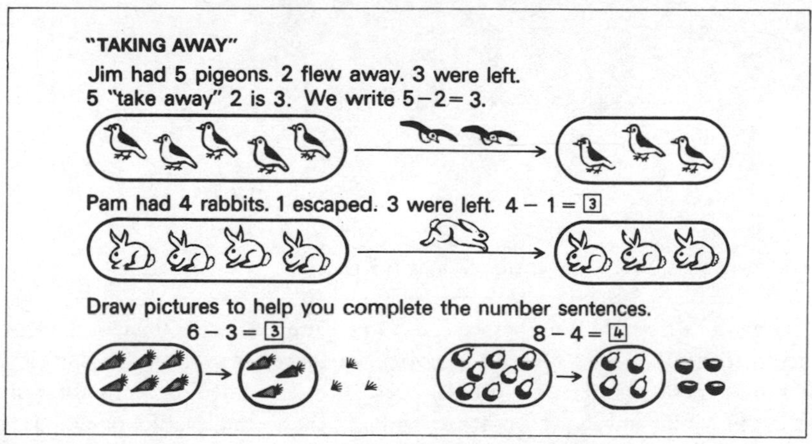

Figure 5.19 Use of motion to indicate 'taking away'.

that some of the carrots and acorns have been removed. However, reading problems may be caused by the picture of two birds apparently flying away towards the three which remain, and by the picture of the escaping rabbit apparently running towards the three which are left. The pictures of carrot-tops and acorn cups may also be difficult to interpret.

The *Bronto Books* (1979), issued in conjunction with the *Nuffield Maths 5–11* series, are intended to help young children to read and understand mathematical vocabulary. The two pages of a Bronto book shown in Figure 5.20 make use of conventional cues to indicate motion. They are intended to give the child cues to the meaning of the words 'round and round' and 'down'.

Figure 5.20 Conventional cues for motion.

The pictures from the same series shown in Figure 5.21 are intended to encourage the child to try to make a one-to-one correspondence between the presents and the little birds. These pictures use postural cues. The arrival of Frog in the first picture is indicated by his unbalanced posture, while in the second picture one of the baby birds is shown in motion as he leans forward to lift his scarf out of the box.

Figure 5.21 Postural cues for motion.

The use of symbols in diagrams

Symbols are often used in diagrams, just as they are often used in mathematical text. Some symbols are common to diagrams and text, while others have been specially devised for use in diagrams alone. For example, in a diagram of a solid, a broken line is often drawn to indicate an edge which would be invisible from the viewpoint from which the solid is drawn. Children often have difficulty both in interpreting and drawing diagrams of three-dimensional shapes, and help may be needed to enable them to visualise what is meant. Figure 5.22 is an extract from *Modern Mathematics for Schools*, Book 1 (1971). This exercise is intended to help the pupils, 12-year-olds of above-average ability, to read and draw diagrams of solid figures.

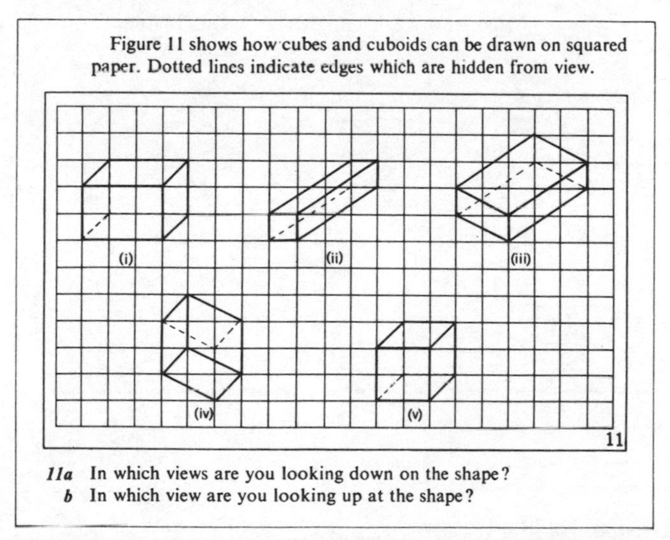
Figure 11 shows how cubes and cuboids can be drawn on squared paper. Dotted lines indicate edges which are hidden from view.

11a In which views are you looking down on the shape?
b In which view are you looking up at the shape?

Figure 5.22 A lesson in reading and drawing a diagram.

The use of the multi-purpose arrow symbol to denote a relation has been discussed in Chapter 4. There are many other ideas which are often symbolised by arrows, in addition to relations. An example of the varied uses of arrows is shown in

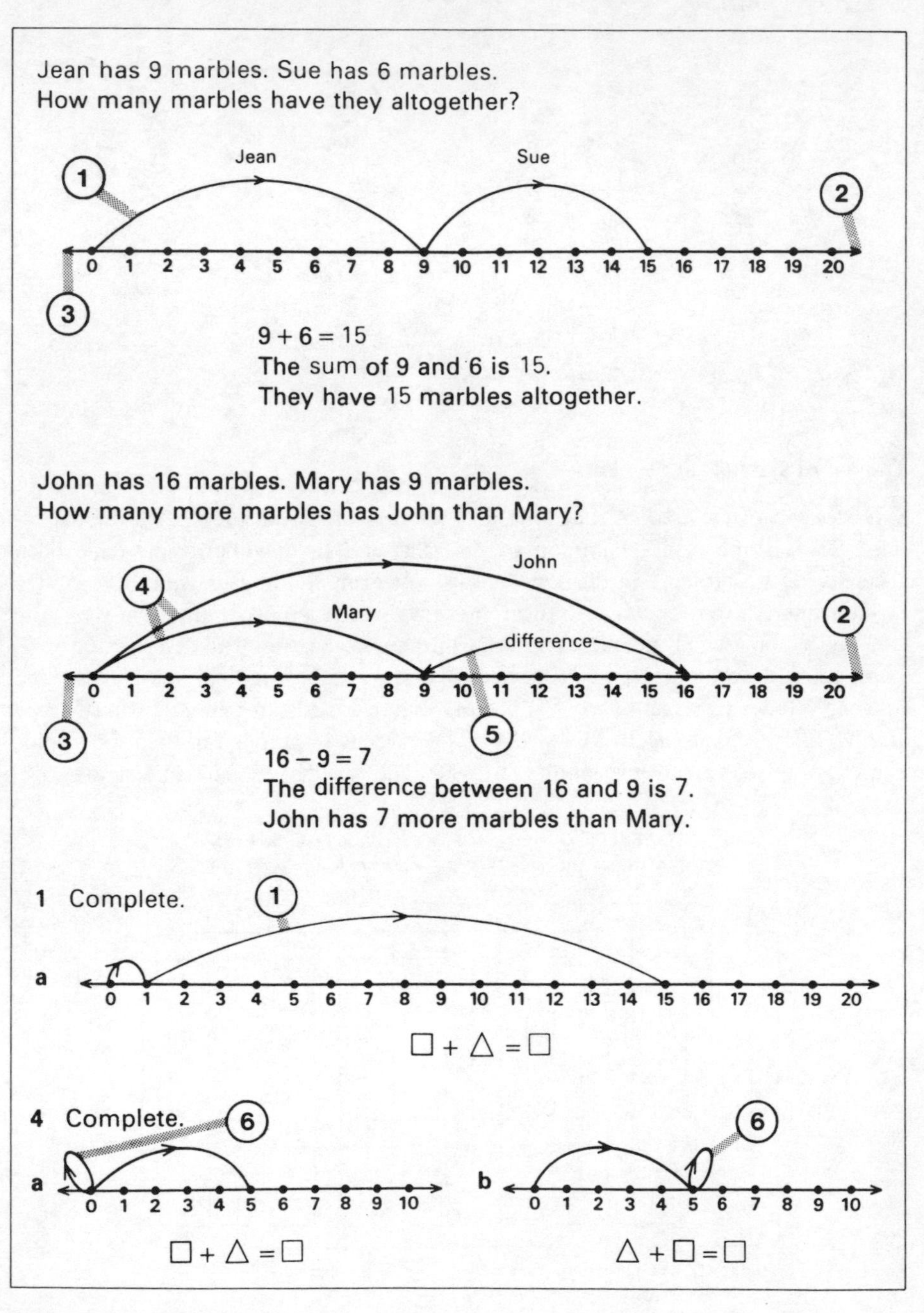

Figure 5.23 Six uses of an arrow.

Figure 5.23. These diagrams all appear in the same section of *Mathematics for Schools,* Level II, Book 1, second edition (1980). The diagrams contain six different uses of an arrow, none of which denotes a relation. The uses are listed below:

1. Arrows show successive jumps along the number line, and are intended to help the child to add numbers.
2. This arrow indicates that the number line can be continued indefinitely to show larger and larger numbers.
3. This arrow symbolises the presence of numbers to the left of zero. It is probably wasted on most children at this stage.
4. The jumps shown by these arrows start from the same point, and are not successive. They symbolise two numbers whose difference is to be found.
5. This arrow has arrowheads at both ends. It does not represent a jump, but marks out the difference between the two jumps.
6. This 'zero jump' arrow conventionally indicates the addition (or subtraction) of zero, which could not otherwise be shown on a number line.

In order for the seven-year-old pupils for whom this text is intended to interpret these diagrams, they have to understand the conventions used, and this requires detailed discussion with the teacher.

We have seen in this chapter that the pictures and diagrams used in mathematical text have their own conventions, which pupils need to learn to read. If a child does not understand all that is implied by an illustration in one of his reading books, his grasp of the story may not be substantially impaired. However, if he cannot read the graphic language of his mathematics book as well as the words and symbols, he will not be able to get the full message from the page.

6 The flow of meaning

Introduction

In addition to decoding the words, symbols and graphic material in mathematical text, the reader needs to study it at a second, and more important level. He needs to make the ideas and argument his own. The purpose of exposition in mathematical text is to enable the reader to reach some new understanding of mathematics through his reading. If this is to happen, the text must have a clear 'story line' or *flow of meaning* as the argument proceeds through the page or chapter.

Many authors of mathematical text for children wish their texts to be used as a framework for learning by guided discovery. Consequently, as a matter of policy, these authors do not explicitly state all the steps in the argument. They prefer to provide activities or questions through which the reader (it is hoped) will discover the ideas for himself in an active and meaningful way. This method of exposition is excellent when it works, but if the pupil does not discover what was intended, the purpose of the text is lost. The purpose is also lost if pupils do not understand those ideas which are explicitly stated in the text.

When the teacher uses a teaching style in which he expects pupils to learn mathematics through reading, it is important for the meaning to flow as clearly as possible through the text, so that pupils acquire the meaning easily from their reading. It is also important that the teacher should identify those steps in the argument which the pupil is expected to discover for himself from the questions and activities; the teacher can then verify by questioning that the pupils have in fact attained the intended meanings.

Flow-of-meaning diagrams

In order to help us to analyse the flow of meaning in text, we have adapted a technique for analysing text used by the Schools Council project *Reading for Learning in the Secondary School* (1980). They suggest that a text should be segmented into *meaning units*; a flow diagram can then be drawn to make clear the flow of meaning through the argument. In drawing such flow diagrams, we have used different symbols to distinguish between those statements which are explicitly made in the text, and those which the pupil is meant to infer from questions and activities he is asked to do. In the course of this analysis, we became aware of a third type of meaning unit: there are some meanings which the pupil needs to infer from the text, but his attention is not explicitly drawn to these meanings by any of the questions or activities: the author has relied on the fact that the pupil will be able to supply these meanings for himself. Meaning units of this third type also include ideas which are not stated in the text, but must be called upon from the reader's previous knowledge in order to supply an essential step in the argument. Often meaning units of this third type are generalisations or ideas which are important for the understanding of later parts of the argument.

Thus, in drawing flow diagrams to show the flow of meaning through an argument, we have used symbols for three types of meaning unit, as follows:

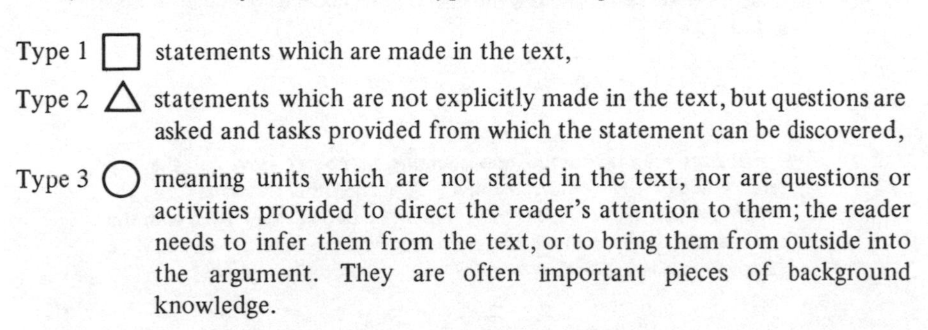

Type 1 □ statements which are made in the text,

Type 2 △ statements which are not explicitly made in the text, but questions are asked and tasks provided from which the statement can be discovered,

Type 3 ○ meaning units which are not stated in the text, nor are questions or activities provided to direct the reader's attention to them; the reader needs to infer them from the text, or to bring them from outside into the argument. They are often important pieces of background knowledge.

A flow-of-meaning diagram is a technique for looking at text itself, irrespective of what the reader brings to the text. Further analysis is necessary to assess the effect on the meaning of the information the reader himself brings to the reading of the text. Thus, in a flow-of-meaning analysis, we regard a meaning unit as of Type 1 if it is explicitly stated, whether or not a particular reader is likely to understand it. For example, the statement:

> In measuring the dimensions of a small rigid object, such as a machine part being turned on a lathe, vernier callipers or a micrometer screw gauge may be used.
>
> (*SMP New Book 3*, 1981)

is a Type 1 statement, even if none of the pupils in a particular group who are reading the text has seen or heard of vernier callipers or a micrometer screw gauge.

We give two examples of flow-of-meaning diagrams. These call attention to the structure of the argument in the exposition studied. When pupils encounter a difficult passage of text, the teacher may find it useful to draw a flow-of-meaning diagram as an aid in analysing the difficulties. He may then be able to apprise pupils of meaning units which are missing from the text itself.

A passage from *SMP Book B*

We shall discuss the first part of the passage in *SMP Book B*, Chapter 5 (1979), which contains introductory work on fractions (Figure 6.1, p. 68). This passage was intended to provide a fresh presentation of fractions to 11/12-year-old pupils of average ability who have already done some work on fractions at their primary schools. Many pupils find the passage difficult to comprehend. Figure 6.2 shows the construction of a flow-of-meaning diagram for this passage. The passage is intended for use as a framework for discovery learning, and consequently it contains many

1. REPRESENTING FRACTIONS

(*a*) In Book A, we showed the counting numbers as points on a line, as in Figure 1.

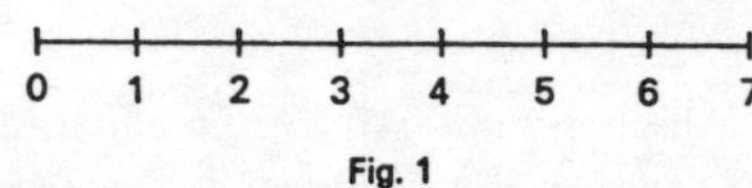

Fig. 1

Fractions are made up of two counting numbers? How can they be represented? How would *you* show $\frac{1}{4}$, $\frac{1}{2}$, $\frac{3}{4}$ and other fractions?

Perhaps you might decide to show them on two number lines, with the top number on one and the bottom number on the other. They could be joined with a line, like this

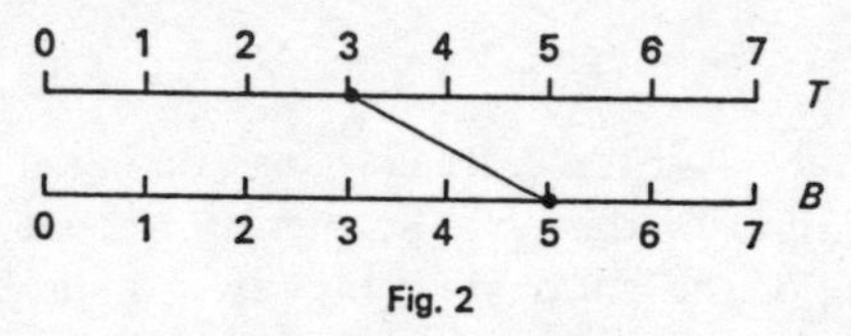

Fig. 2

What fraction do you think is shown in Figure 2?

Can you think of a better position for the *T* and *B* number lines than having them one above the other?

(*b*) Try drawing the *B* line at right-angles to the *T* line, meeting at 0, as in Figure 3.

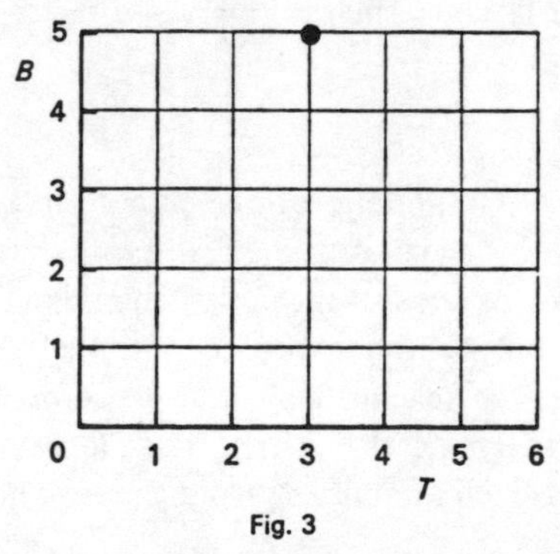

Fig. 3

Figure 3 shows $\frac{3}{5}$ again, but in a much neater way than is shown in Figure 2. Where do you remember seeing a different kind of number pair plotted in this way? We shall refer to this way of representing fractions as *graphing* the fractions.

Figure 6.1 The text of the passage.

1. REPRESENTING FRACTIONS

[1] (*a*) In Book A, we showed the counting numbers as points on a line, as in Figure 1.

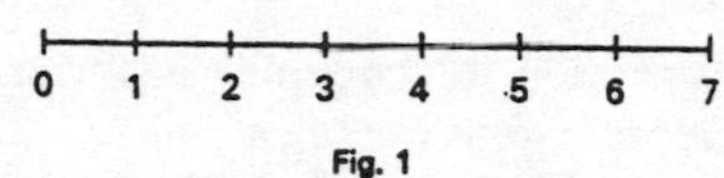

Fig. 1

[2] Fractions are made up of two counting numbers? How can they be represented? How would *you* show $\frac{1}{4}$, $\frac{1}{2}$, $\frac{3}{4}$ and other fractions?

(A_1) Fractions can be shown on the number line between the points representing integers.

(A_2) Fractions can be shown by shading parts of rectangles or circles.

(B) Fractions are made up of *number pairs* of counting numbers.

Perhaps you might decide to show them on two number lines, with the top number on one and the bottom number on the other. They could be joined with a line, like this

[3]

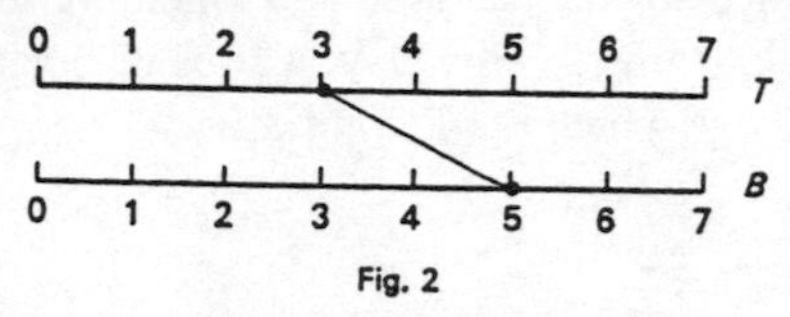

Fig. 2

What fraction do you think is shown in Figure 2?

[4] Can you think of a better position for the *T* and *B* number lines than having them one above the other?

[5] (*b*) Try drawing the *B* line at right-angles to the *T* line, meeting at 0, as in Figure 3.

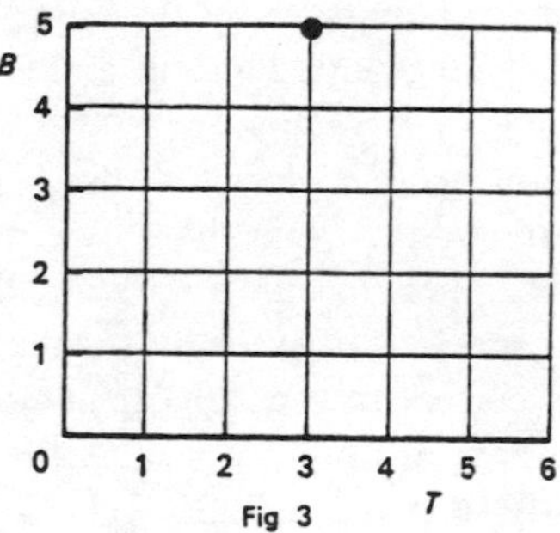

Fig 3

Figure 3 shows $\frac{3}{5}$ again, but in a much neater way than is shown in Figure 2.

[6] Where do you remember seeing a different kind of number pair plotted in this way?

(C) Number pairs can be graphed.

[7] We shall refer to this way of representing fractions as *graphing* the fractions.

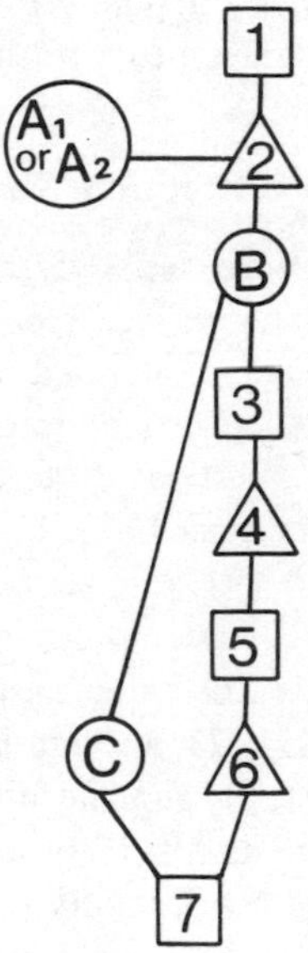

Figure 6.2 *left* Segmentation into meaning units. *right* The flow-of-meaning diagram.

questions. One of its most interesting features, however, is the presence of 'red herrings'. The question:

How would *you* show $\frac{1}{4}$, $\frac{1}{2}$, $\frac{3}{4}$ and other fractions?

is very likely to point the pupils' attention in the wrong direction, as their only previous knowledge of methods of representing fractions will be of one or both of those shown in the Type 3 meaning units (A1) and (A2) (Figure 6.2), rather than the graphical method which is developed later in the text. The argument which then follows depends on the Type 3 meaning unit (B), which was hinted at in △2, although 'number pairs' were not then mentioned. The diagram which is part of [3] shows the points marked on parallel lines, and so leads the reader towards the graph whose construction is the object of the passage. However, this first diagram is also a red herring, because only the final graph is to be used later in the passage. The question in △4 is a rhetorical question, which invites the answer 'No'.

In Chapter 9, we shall discuss the construction of a simplified version of this passage of text, obtained by omitting the red herrings and the rhetorical questions which spoil the flow of meaning through the passage. This amended version proved easier for pupils to read and work from. In this way, a flow-of-meaning diagram provides an aid to simplifying a passage which pupils find difficult.

An extract from *SMP New Book 3,* Part 1

A more complicated flow-of-meaning diagram is needed to understand the structure of the passage from *SMP New Book 3* (1981) which will now be discussed. This book is intended for 13–14-year-old pupils of above average ability who use calculators as an important tool in their mathematics. The passage focusses attention on questions of accuracy which are made important by the use of calculators. The text is shown in Figure 6.3, and has been segmented into meaning units (numbered 1 to 19) in Figure 6.4, pp. 72 and 73.

An attempt to draw a flow diagram made it clear that much of the argument rested on three unstated Type 3 meaning units, which have been inserted in Figure 6.4, p. 73, and are labelled A, B and C. These meaning units are all generalisations; first, the pupil needs to retain from the previous page the meaning of a measurement such as $L = 12.6$ cm, when it is given correct to a particular number of significant figures. Secondly, he needs to know how to multiply inequalities, and finally, he must move easily between an inequality such as $112.3225 \leqslant A < 114.4825$ and the alternative form 113.4 ± 1.1.

Figure 6.4, p. 73, shows a possible flow diagram for the meaning of the passage; this indicates the crucial importance of the three Type 3 meaning units, on which most of the argument depends.

We notice that this passage contains only one Type 2 meaning unit:

△7 Is it then correct to give the area as $113.4\ \text{cm}^2$?

There is little for the pupils to discover for themselves in this passage, which is a rather complicated argument expounded through an example. The structure of the

2. COMBINING MEASUREMENTS

Measurements often have to be combined in some way: for example, an average speed may be calculated by dividing a distance measurement by a time measurement, or the perimeter of a field may be calculated by adding together several different measurements of lengths along its boundaries. The final results of such calculations can never be known to greater accuracy than the least accurate of the measurements involved.

If the length. L cm, of a piece of paper is recorded as 12.6 cm it means that $12.55 \leqslant L < 12.65$. If the width, W cm. is recorded as 9.0 cm it means that $8.95 \leqslant W < 9.05$. If the area ($L \times W$ cm^2) is now calculated as 12.6×9.0 the result is 113.4. Is it then correct to give the area as 113.4 cm^2?

In fact the area, A cm^2, cannot be known for certain to this accuracy. What is certain is that area cannot be smaller than 8.95×12.55 cm^2 and that it must be less than 9.05×12.65 cm^2.

i.e.
$$8.95 \times 12.55 \leqslant A < 9.05 \times 12.65$$
$$112.3225 \leqslant A < 114.4825$$

112.3225 is known as the 'lower bound' for the value of A;
114.4825 is known as the 'upper bound' for the value of A.
It would therefore be misleading to give the area as 113.4 cm^2 since that would suggest that $113.35 \leqslant A < 113.45$, and we cannot be sure of this.

We could, however, use the bounds to write the area as 113.4 ± 1.1 cm^2.

In practical problems involving calculations from measurements it would be a nuisance to have to calculate the upper and lower bounds every time. We shall adopt a convention that an answer will be given to the same number of significant figures as the least accurate of the original measurements. But we shall need to examine from time to time whether this convention is reasonable for a particular calculation. The only sure way of knowing the accuracy of an answer is to calculate the upper and lower bounds.

In the above example the measurement of the width was given as 9.0 cm, to 2 s.f. The area, calculated as 113.4 cm^2, would therefore be given as '110 cm^2 to two significant figures'. We are claiming that

$$105 \leqslant A < 115$$

and this is certainly true.

Figure 6.3 The text of the passage.

argument in the flow diagram shows that the first four meaning units form an introduction. The main argument begins with statement [5]; this uses the Type 3 generalisation (A) which the pupil is expected to have made earlier. A formal statement of this generalisation (the meaning of a measurement which is given to a particular number of significant figures) has been inserted at (A), although the pupil will use it informally in examples only; its important role in the argument is clear. After $L \times W$ has been calculated, the pupil is expected to know how inequalities are multiplied, so another Type 3 unit has been inserted at (B), in

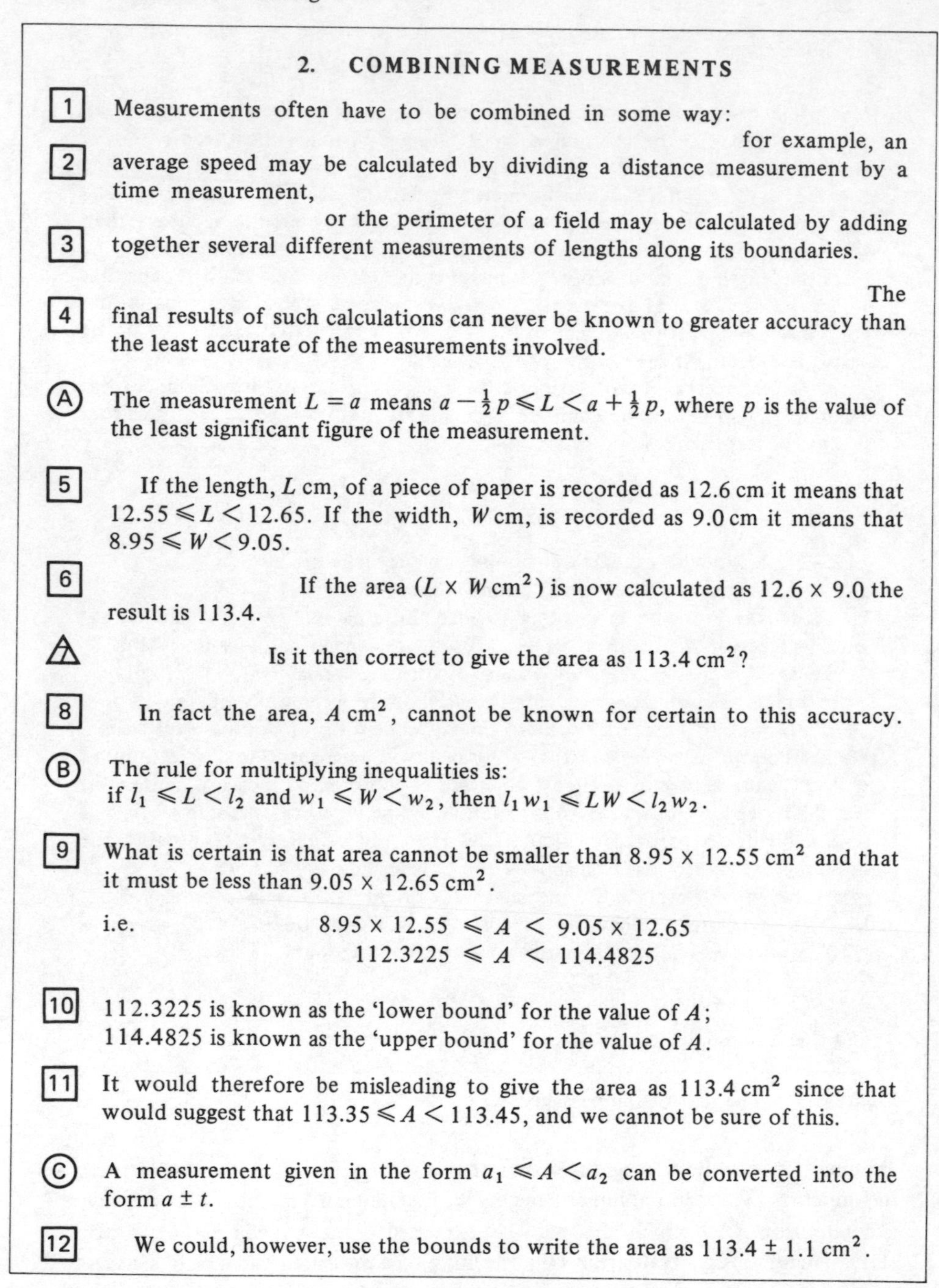

2. COMBINING MEASUREMENTS

1 Measurements often have to be combined in some way:

2 for example, an average speed may be calculated by dividing a distance measurement by a time measurement,

3 or the perimeter of a field may be calculated by adding together several different measurements of lengths along its boundaries.

4 The final results of such calculations can never be known to greater accuracy than the least accurate of the measurements involved.

(A) The measurement $L = a$ means $a - \frac{1}{2}p \leqslant L < a + \frac{1}{2}p$, where p is the value of the least significant figure of the measurement.

5 If the length, L cm, of a piece of paper is recorded as 12.6 cm it means that $12.55 \leqslant L < 12.65$. If the width, W cm, is recorded as 9.0 cm it means that $8.95 \leqslant W < 9.05$.

6 If the area ($L \times W\ \text{cm}^2$) is now calculated as 12.6×9.0 the result is 113.4.

7 Is it then correct to give the area as 113.4 cm^2?

8 In fact the area, $A\ \text{cm}^2$, cannot be known for certain to this accuracy.

(B) The rule for multiplying inequalities is:
if $l_1 \leqslant L < l_2$ and $w_1 \leqslant W < w_2$, then $l_1 w_1 \leqslant LW < l_2 w_2$.

9 What is certain is that area cannot be smaller than $8.95 \times 12.55\ \text{cm}^2$ and that it must be less than $9.05 \times 12.65\ \text{cm}^2$.

i.e.

$$8.95 \times 12.55 \leqslant A < 9.05 \times 12.65$$
$$112.3225 \leqslant A < 114.4825$$

10 112.3225 is known as the 'lower bound' for the value of A;
114.4825 is known as the 'upper bound' for the value of A.

11 It would therefore be misleading to give the area as 113.4 cm^2 since that would suggest that $113.35 \leqslant A < 113.45$, and we cannot be sure of this.

(C) A measurement given in the form $a_1 \leqslant A < a_2$ can be converted into the form $a \pm t$.

12 We could, however, use the bounds to write the area as $113.4 \pm 1.1\ \text{cm}^2$.

Figure 6.4 *above and upper right* Segmentation into meaning units. *lower right* Flow-of-meaning diagram.

13 In practical problems involving calculations from measurements it would be a nuisance to have to calculate the upper and lower bounds every time.

14 We shall adopt a convention that an answer will be given to the same number of significant figures as the least accurate of the original measurements.

But

15 we shall need to examine from time to time whether this convention is reasonable for a particular calculation.

16 The only sure way of knowing the accuracy of an answer is to calculate the upper and lower bounds.

17 In the above example the measurement of the width was given as 9.0 cm, to 2 s.f.

18 The area, calculated as $113.4\ \text{cm}^2$, would therefore be given as '$110\ \text{cm}^2$ to two significant figures'.

We are claiming that

$$105 \leqslant A < 115$$

19 and this is certainly true.

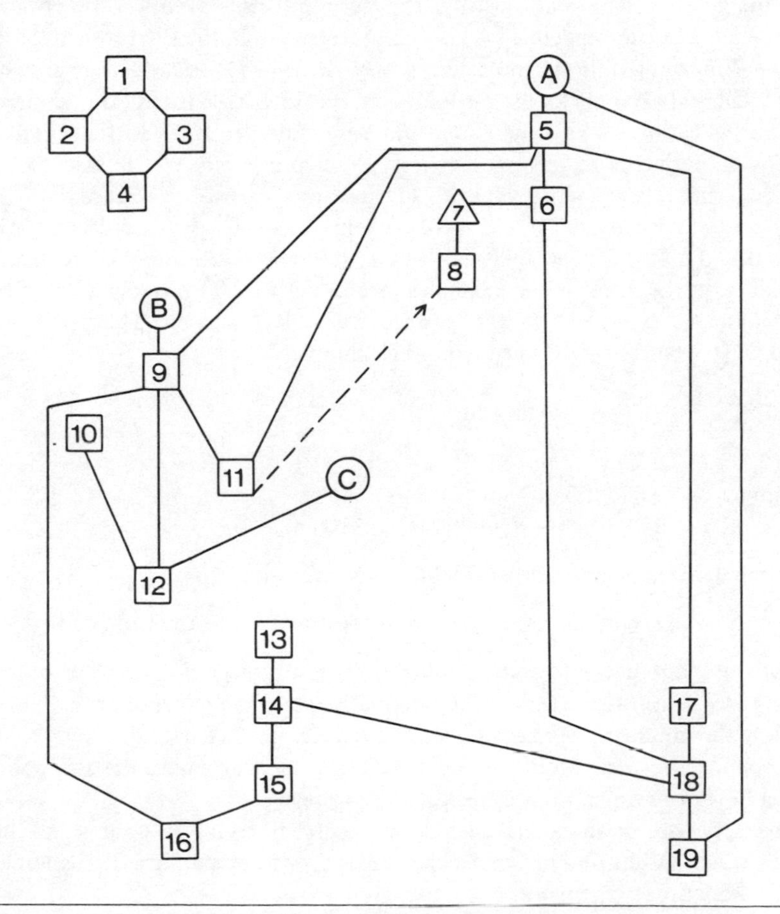

order to express this. Pupils would probably state the idea expressed in (B) in the alternative and more pictorial form 'a rectangle with smaller dimensions has a smaller area'. The statement has been expressed in inequality form because this form is used in [9]. At the next step in the argument, statement [11] is:

It would be misleading to give the area as 113.4 cm^2

This again relates back to (A), so that the reader has to keep in mind throughout the passage what is meant by giving a measurement to a particular degree of accuracy. The first part of the example ends with the conversion of the inequality for area into the form 113.4 ± 1.1 cm^2, which the pupil may or may not know a technique for doing. Hence, the Type 3 meaning unit (C) has been inserted to express the need for this technique.

A new theme is introduced in [14], when a convention is stated:

Answers will be given to the same number of significant figures as the least accurate of the original measurements.

The remainder of the passage justifies the fact that this convention will always give a true result. The meaning unit (A) is again referred to in the final statement [19].

Type 3 meaning units are not always easy to identify, because they are implied rather than stated; a slight discontinuity in the smooth flow of the argument is often an indication that a Type 3 unit is lurking underneath. Type 3 units are often intended to be in the pupil's mind as a result of previous work. In this case, statement (A) and a statement related to (C) are the focus of the preceding page of text (shown in Figure 6.5). A flow-of-meaning analysis of this preceding passage shows that (A) is a generalisation which the pupils are intended to make for themselves on the basis of the examples given, and so (A) is a Type 3 unit in this earlier passage, as well as in the passage we have discussed. The generalisation related to (C) has to be obtained from the examples,

0.520 ± 0.001 cm means 0.519 cm $\leqslant$ length $\leqslant$ 0.521 cm

and

Diameter = 0.825 ± 0.002 cm means
0.823 cm $\leqslant$ diameter $\leqslant$ 0.827 cm.

This generalisation is not in fact (C), but its converse:

(C′) A quantity in the form $a \pm d$ can be converted into the form $a_1 \leqslant A \leqslant a_2$.

Pupils are not likely to notice many of the subtle points which the flow-of-meaning diagram brings out, but they may experience a feeling of general inability to follow the argument of the passage. The teacher needs to decide how far, in preparing pupils for their work on these passages, he will emphasise the points of which a flow-of-meaning analysis makes him aware.

Thus, a flow-of-meaning analysis may be useful in trying to understand the difficulties which pupils find in a particular passage, or in preparing a discussion lesson as an introduction to the work.

Specifying measurements

In measuring the dimensions of a small rigid object, such as a machine part being turned on a lathe, vernier callipers or a micrometer screw gauge may be used. If the diameter of a metal rod is given as 0.362 cm it means that as far as can be judged from the scale, the diameter is nearer to 0.362 cm than to 0.361 cm or 0.363 cm, i.e. that

$$0.3615\text{ cm} \leqslant \text{diameter} < 0.3625\text{ cm}.$$

(If it is possible to decide between two readings, there are a number of possible actions. One possibility is always to give the higher number; another is to give the higher or lower in such a way that the last digit is even. We have assumed that the former course of action has been taken.)

When someone manufacturing a part for a machine reads his drawing, he may find a width given in a different form:
0.520 ± 0.001 cm, say. This means:

$$0.519\text{ cm} \leqslant \text{length} \leqslant 0.521\text{ cm}$$

and shows how much room for error he can allow himself. The ± 0.001 cm is known as the 'tolerance'.

Here are some further examples of these inequalities:

(1) Current = 1.5 A means

$$1.45\text{ A} \leqslant \text{current} < 1.55\text{ A}.$$

(2) Temperature = 273.0 K means

$$272.95\text{ K} \leqslant \text{temperature} < 273.05\text{ K}.$$

But temperature = 273 K means

$$272.5\text{ K} \leqslant \text{temperature} < 273.5\text{ K}.$$

(3) Diameter = 0.825 ± 0.002 cm means

$$0.823\text{ cm} \leqslant \text{diameter} \leqslant 0.827\text{ cm}.$$

(4) Resistivity = 2.5×10^{-6} Ω m means

$$2.45 \times 10^{-6}\ \Omega\text{ m} \leqslant \text{resistivity} < 2.55 \times 10^{-6}\ \Omega\text{ m}.$$

(5) Length = 250 m to 2 s.f. means

$$245\text{ m} \leqslant \text{length} < 255\text{ m}.$$

Figure 6.5 The passage of exposition preceding the text of Figure 6.3.

7 Readability formulae and their limitations

Factors which affect readability

Throughout this book we have used the word *readability* to convey the idea that in a readable text it is easy for the reader to *get the meaning from the page.*

The two passages shown in Figure 7.1, a passage of OE by Ivan Illich and a passage of ME taken from *New General Mathematics Revision* by Channon, McLeish

Teachers just cannot be blamed for the failures of a revolutionary government that insists on the institutional capitalization of manpower through a hidden curriculum guaranteed to produce a universal bourgeoisie.

Example 4 $y \propto x^2$. *How is the value of y affected if the value of x decreases by 20%?*

Let $y = kx^2$ (i)

Let y become y' when x decreases to $\frac{80}{100}x$.

Then $y' = k\left(\frac{80}{100}x\right)^2$ (ii)

Dividing (ii) by (i),

$$\frac{y'}{y} = \frac{k\left(\frac{80}{100}x\right)^2}{kx^2} = \left(\frac{80}{100}\right)^2 = \frac{64}{100}$$

$$y' = \frac{64}{100}y$$

y is decreased by 36%.

Figure 7.1 *above,* Illich (1974); *below,* Channon, McLeish Smith and Head (1976).

Smith and Head, are both difficult to read. There are many reasons for this. In both passages we have to decide what the author means by various words, phrases and symbols. But even when we know the separate meanings of these units, we have to try to grasp the whole. Not only do the individual words and symbols affect the readability, but so do a host of other factors, such as the layout of the page, the previous knowledge which the reader is expected to bring to the passage, and the logical steps which the author misses out of the argument because he assumes them to be obvious.

Readability formulae

Many readability formulae have been devised to help in assessing the readability of OE text. Typically they involve measuring some aspects of the text, and relating these aspects to the ease with which the user reads the text. We need to consider whether readability formulae can be applied to ME texts as well as to OE texts. In fact, almost every readability formula has been developed on the basis of a study of continuous English prose, whereas mathematics books for children often contain very little continuous prose. Moreover, almost all the formulae derive from American studies and so are based on American English usage. They may or may not apply directly to British materials. The formulae may also be criticised on other grounds, as we shall see later in this chapter.

We shall look first at a few of the more popular readability formulae and how they are applied to OE texts; we then discuss how such formulae have been applied to mathematics texts, and consider formulae which have been devised especially for mathematical text. Lastly we look at the pitfalls which abound in the area of testing readability.

Readability formulae for continuous prose

The method of devising a readability formula is essentially simple. A number of passages of text are selected which have already been graded for their ease or difficulty of reading. Characteristics of these passages which might appear to affect their readability are listed. The variables which might be studied include average word length, average number of syllables per word, and average number of words per sentence. The number of unfamiliar words in the passage, and many other possible variables, can also be considered. The passages are examined to see which of the variables correlate well with reading difficulty. Finally, a number of variables which correlate best with reading difficulty are combined into a formula.

A few of the more widely used formulae are given below. Most are intended to be used on one or more 100-word passages selected from the text. Many formulae give their results as American grade levels; to obtain a rough chronological age, 5 should be added. Further details of how to apply these and other readability formulae, and a detailed discussion of their strengths and weaknesses, can be found in Harrison (1980).

The Dale-Chall formula

This formula, devised by Dale and Chall (1948), is based largely on unfamiliar words, and takes little account of the syntax or structure of sentences. Dale had previously drawn up a list of 3000 familiar words, and this was used as the basis of the formula, together with sentence length. The formula was originally stated as:

$$C_{50} = 0.1579X + 0.0496W + 3.6365$$

where

X = percentage of words not on the Dale list of 3000 familiar words
W = average number of words per sentence
C_{50} = the reading grade score of a pupil who could answer correctly 50 per cent of the questions on a reading comprehension test covering the passage.

The variable C_{50} may be thought of as the US grade level required to read the passage. In later work, Dale and Chall came to regard their formula as underestimating the level of more difficult materials, and so a revised list of equivalent grade levels was developed. For example, material with a Dale-Chall score between 8 and 9 was then regarded as suitable for 16–17-year-olds rather than 13–14-year-olds.

The Flesch Reading Ease formula
This formula, devised by Flesch (1948), uses the lengths of words and sentences. It was originally based on 'thought units' rather than sentences, but since it is difficult to agree on what constitutes a 'thought unit', this idea had to be dropped. The Reading Ease is an index number ranging from 100 (easy) to 0 (difficult). The formula is:

$$\text{Reading Ease} = 206.835 - 0.846S - 1.015W$$

where

S = average number of syllables per 100 words
W = average number of words per sentence

The Reading Ease index can be converted to a grade level by using either a nomogram or one of several algorithms chosen according to the actual score and taken from the STAR computer program devised by General Motors (Harrison, 1980). For example, if the Reading Ease is over 70, the following formula is used:

$$\text{Grade level} = -[(\text{Reading Ease} - 150)/10]$$

Gunning's FOG formula
This formula (Gunning, 1952) owes much of its popularity to the fact that it is very easy to use. It is

$$F = 0.4(W + P)$$

where

W = average number of words per sentence
P = percentage of words containing 3 or more syllables (excluding those ending in *-ed* or *-ing*)
F = US grade level

Gunning used the acronym FOG to stand for 'frequency of gobbledegook'!

The Fry Readability Graph

All the above formulae involve an amount of computation which even a mathematics teacher may find inconvenient.

Edward Fry developed a Readability Graph which is simpler to use than any of the formulae; it employs the same two measures of difficulty of text as the Flesch formula, but does not require any calculations. Fry's Readability Graph was first developed for use in Uganda, and was published in Britain in 1961. Its appearance in revised form in America in 1968 and 1969 (Fry, 1969), geared to American

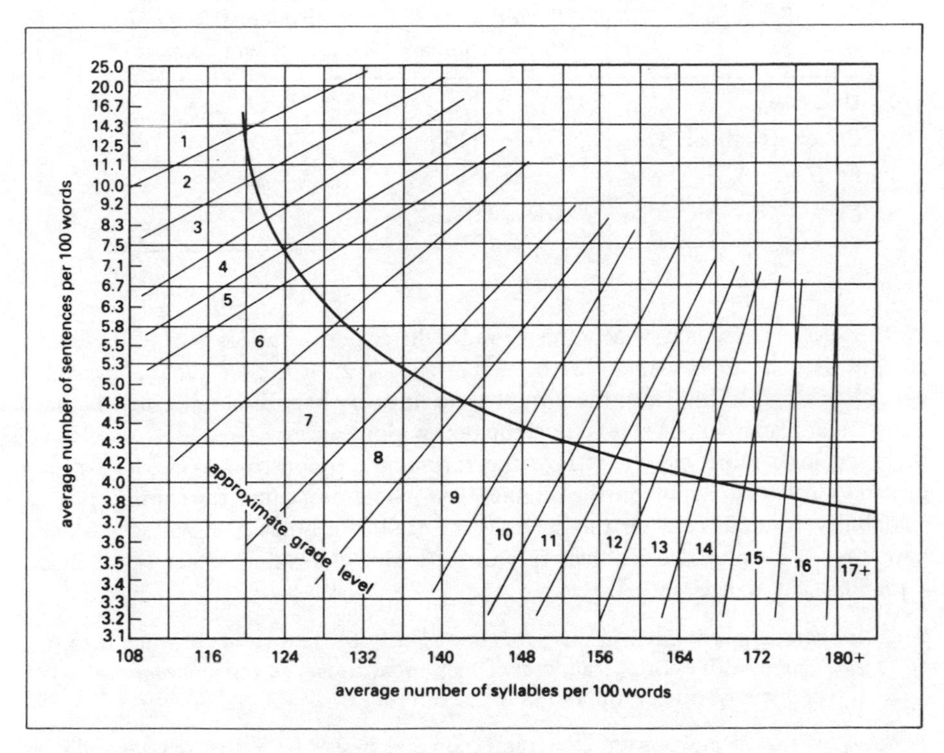

Figure 7.2 The Fry Readability Graph (after Harrison, *Readability in the Classroom*, 1980).

grade levels, made it a widely used tool. The graph is reproduced in Figure 7.2 in a form which gives the UK reading age. Directions for its use are:

1. Select three 100-word passages from the beginning, middle and end of the book.
2. Count the number of sentences in each passage. Find the average.
3. Count the number of syllables in each passage. Find the average.
4. Plot the average number of sentences against the average number of syllables on the graph of Figure 7.2 and read off the reading age.

Criticism of readability formulae

Different readability formulae may give very different estimates of the readability of the same passage. To show this variability, each of the above formulae has been used on the first paragraph of the section in this chapter entitled 'Readability formulae for continuous prose', p. 77. Since the passage is a short one, some extrapolation is necessary.

Table 7.1

Formula	Result given by the formula	Equivalent UK reading age (where possible)
Dale-Chall	C_{50} = 9.6	College
Flesch Reading Ease	Grade = 13	18
FOG	Grade = 17	22
Fry	UK age = 16	16

The results show how widely the formulae differ in their assessment of the same passage of text: FOG claims that a reading age of 22 is needed for this passage, while Fry suggests 16. Thus, the formulae are not very helpful in assessing whether a particular reader will be able to read the text with meaning.

If we look more closely at how the formulae are constructed, we see other factors which cast doubt on their value. The whole procedure for constructing a readability formula rests on the necessity of finding a number of passages which have already been graded for their ease or difficulty of reading – that is, for their readability. As Manzo (1976) says:

> If there is any validity in this process it is only to the extent to which there is agreement with existing standards. This is incestuous . . . and makes readability research a construct without a point of reference.

Many of the formulae were constructed on the basis of a set of passages which were originally graded in the 1920s by a group of librarians and teachers (Crabbs and McCall, 1926). So the formulae merely correlate the textual variables with the judgments of a group of reading experts; not even a measure such as the relative success of groups of children in comprehending the passages was used in devising these formulae. If we look at the dates when the formulae were devised we may also feel some concern; a formula devised in 1948 may still be applicable, but it is quite likely that both children's experience and the texts they are accustomed to reading have changed in the past thirty years. Indeed, the original McCall-Crabbs passages were graded more than half a century ago.

Most of the formulae originated in the United States, and are based on American experience. The Dale-Chall list of the 3000 words best known by American 8-year-olds was compiled in 1948 and omits some words which were amongst those most commonly used by English 7-year-olds at a later date; these words include *bonfire, mummy* and *television* (Edwards and Gibbon, 1964, 1973).

Another important criticism is that the formulae only use variables which can be easily quantified; any feature of text that cannot be given a numerical value must be ignored. Clearly, the syntax of a sentence greatly influences its readability, but the only measure which reflects syntax in most of the formulae is sentence length. A number of studies, such as those of Reid (1972), Bormuth *et al.* (1970), Van der Will (1976), have shown that 'shorter' does not necessarily mean 'simpler'. For example, we may compare

Mary's dress was neither new nor pretty.

with

Mary's dress was not new and it was not pretty.

Subjects scored over 43 per cent more on comprehension tests in the longer sentence (Reid, 1972).

Readability formulae also necessarily ignore the *legibility* of the text; legibility is based on visual factors such as the typeface, the length of the lines and the layout of the pages. We shall see in Chapter 8 that all these factors, which make the text more or less legible, have a considerable influence on readability.

As we have seen, readability formulae are also unreliable; different formulae do not give the same results on the same texts. More importantly, they do not differ consistently. Stokes (1978) states:

> Many of those using readability formulae will be doing so in order to collect more information to assist them in determining whether or not a text is appropriate for a particular readership. Teachers in particular will be interested in knowing the age range with which a text may be used. It is in this respect that readability formulae are at their least useful – indeed, they may be positively misleading. The formulae do not give appreciably similar results even in the long run and . . . cannot be standardised by conversion factors. To assert that formulae are nevertheless very good at rank-ordering passages is in this context largely irrelevant.

Of course, different passages of a single text may vary considerably in their readability, especially if more than one author has contributed to the book.

Moreover, readability formulae ignore the influence of pupil motivation. Many pupils read and enjoy exciting novels which have a reading difficulty well above their apparent capabilities. The matching of the reader to the text has of necessity to be ignored. This was recognised by Taylor (1953), who pointed out that:

> . . . readability formulae deal with only one side of the matching exercise, namely the book.

They fail to reflect the effect of the book upon the reader. Taylor, however, was not content merely to criticise; in his 1953 article 'Cloze procedure: a new tool for measuring readability', he put forward a new procedure which was adapted from the old-established technique of sentence completion.

The cloze procedure

The cloze procedure takes its name from a way in which the term 'closure' is used by psychologists; to quote Moyle (1968),

> Gestalt psychologists applied the term 'closure' to the tendency to complete a pattern which has a part missing . . . the fluent reader will often substitute a word of similar spelling and meaning to the one in the text and read correctly a word from which a letter has been omitted.

In the cloze procedure, some words of a text are deleted and replaced by blanks. Readers are asked to supply the missing words. Thus, the cloze procedure measures the interaction between the reader and the text – a reader who finds the text easy will be able to supply more of the missing words than one who does not understand the passage. Typically, every fifth, seventh or tenth word is deleted. Since Taylor first formulated the cloze procedure in 1953, much research has been conducted on it, largely in the USA, and in 1967 Bormuth provided a frame of reference by which to judge particular cloze scores. He compared cloze scores with scores on multiple choice questions, and his results led him to propose a minimum *instructional level* for a text which is to be studied when there is a teacher available to help. When the average percentage score on cloze tests falls below the instructional level, which was set at 40 per cent, then the written material should be considered too difficult for the groups of pupils being tested. A cloze score of 55–60 per cent has been proposed as indicating an *independent level* of readability.

Gilliland (1972) considers that the cloze procedure offers a number of advantages over other measures of readability:

- . . . it appears to reflect the sum total of all influences which interact to affect readability,
- . . . the performance of the reader is being measured on samples of the text to be read,
- When the cloze test is applied, both reader and book are assessed simultaneously by use of the one measure. This undoubtedly gives this procedure a greater face validity than other procedures referred to.

Because features such as type style, line length and layout affect readability (see Chapter 8), it would seem important that the passage containing deletions looks as like the original as possible. Unfortunately, this has not always been the case. In some studies, computers have been used to produce deletions; this may yield text written wholly in capital letters, and it is known that such text is less legible than text written in lower case (see, for instance, Lunzer and Gardner, 1979, p. 89).

The cloze test, when properly used, is a good way to tell if a text matches the reader's abilities. However, to examine the readability of a text by the cloze procedure a group of prospective readers is needed, and these readers may not be available at the time when a school text is being assessed for its suitability for purchase. The cloze procedure is, however, very useful in research on readability. It has been used in studies of such topics as the match between junior school pupils and the mathematics texts they were using. An example of such a study, undertaken by one of the authors, will now be discussed.

A study of the match between junior pupils and their mathematics texts

In 1976, Hubbard (neé Jones) investigated the readability of the mathematics texts in use in junior schools in Sheffield. Her work is described in Jones (1976).

A total of 45 different texts in use in third-year junior classes were scored for readability using the Fry graph. The reading ages obtained ranged from 9 to 15 years, with most of them in the range from 12 years upwards. The cloze procedure was then used to investigate further the match between the most frequently used texts and their readers. A total of 310 third-year junior children in nine schools were used; their mean age was 10 years 3 months, and their mean reading age was 10.78 years, as measured by the GAPADOL test, which uses a cloze procedure to measure reading age. The 310 children also completed cloze tests of three separate 100-word passages from the mathematics books they were actually using in class. The books used were *Alpha 3, Beta 3,* Fletcher's *Mathematics for Schools,* Level II, Books 3, 4 and 5, *Maths Adventure* 3, and Griffiths' *Basic Mathematics.* Among these texts, *Alpha 3* and *Beta 3* predominated, being used in 78 per cent of the schools. The schools used in the study were selected according to the books used, but even so, the children represented a reasonable sample of third-year junior children in Sheffield.

Each 100-word passage was reproduced, omitting every seventh word (and avoiding digits and symbols), as shown in Figure 7.3. The omitted words were, on the whole, 'general' words, which might be encountered in any other book the child might read, and not 'technical' words specific to the field of mathematics.

Graphs (page 6)

Paul and Susan wished to make ——— check of the traffic which passed ——— the busy road near their school. ——— counted the number of motor cycles, ——— cars, buses and lorries which went ——— in a quarter of an hour. Then ——— made this picture.

1. Find the number ———
   a) motor cars
   b) lorries and vans ——— the school.
2. The number of buses ——— less than the number of motor cars.
3. ——— many more lorries and vans than buses passed ———?
4. Find the number of vehicles ———.
5. Suppose one little picture stands for 5 ———, what is then the total of each kind?

Figure 7.3 Passage from *Beta Book 3.*

The passages were reproduced in a form as close as possible to the original books, including pictures and diagrams, and the children were asked to fill in the blanks. In scoring the tests, spelling was not taken into account; the total score for the three passages from each book was combined into a percentage score for that child for that book. Table 7.2 (quoted from Jones, 1976) shows the results.

Table 7.2

Book	Mean cloze score	σ	Fry reading age	% of children below 40%	% of children below 55%	% of children below Fry r.a.
Alpha 3	44.28	14.75	12	35.8	74.8	77.6
Beta 3	48.18	15.73	13	35.8	79.0	85.8
Fletcher Books 3 & 4	43.57	22.13	12	40.6	72.6	76.9
Fletcher Books 4 & 5	46.65	23.57	13	40.4	76.6	85.8
Maths Adventure	41.64	19.29	12	35.7	76	77.6
Griffiths	47.14	19.93	12	52	65	66.6
	1	2	3	4	5	6

It can be seen that there is much closer agreement between columns 5 and 6 of the table than between columns 4 and 6; hence, the reading age given by the Fry Readability Graph is closer to the 'independent level' of reading defined on the cloze procedure by Bormuth, than it is to the 'instructional level'. It has often been argued that reading ages obtained from the Fry Readability Graph are higher than they should be. The results of this study would seem to suggest that the high Fry reading ages are realistic for mathematics texts, which children need to read very closely.

Jones (1954) suggested that three-quarters of an (American) class should be able to read a proposed textbook with ease and understanding. Heddens and Smith (1964) quoted this work and followed it up by suggesting that the reading age of a text should be lower than that of the children with whom the material is to be used. In the mathematics texts investigated by Hubbard, the opposite was the case. Approximately three-quarters of the children could not read the books at the 55 per cent cloze level and one-third could not read them at the 40 per cent 'instructional level'. In all cases, the reading age of the book was above the mean reading age of the children for whom it was designed. In addition, in this study, the mean reading age of the children tested was higher than their chronological age.

If children are expected to learn much of their mathematics from textbooks, they will have problems – not only problems with the mathematics, but problems in reading what they have to do. The information which they gain from reading then has to be translated into a mathematical form, which can be difficult even for a mathematically able child. Thus, the true difficulty of reading a mathematics text is greater than that of merely comprehending the prose. Hubbard's study suggests that teachers should take a hard look at the mathematics texts their children are using; they should consider how to deal with the readability problems of those texts. Only then can the teacher be sure that the children's mathematical progress is not hampered by reading difficulties.

Readability formulae for mathematical text

Because of the differences between mathematical English and ordinary English, the readability tests which have been designed for use on ordinary English text have even more limited application to mathematical English. However, two readability formulae especially designed for school mathematics textbooks have been developed in the USA; these are described in *Helping Children Read Mathematics* by Kane, Byrne and Hater (1974). In reviewing the application of readability measures to school texts in the subject areas, Harrison (1980) pointed out that these specialist mathematics readability measures are almost unknown in Britain, and have not yet been used here.

Kane *et al.* undertook a number of preliminary studies before developing a readability formula. Hater (1969) modified the cloze procedure so that it could be used with mathematical text, and found that the adapted cloze procedure measured the difficulty of passages of mathematical English in a way which correlated well with measures obtained from comprehension tests. Her adaptation involved making a distinction between *word tokens* and *mathematics tokens*. A mathematics token is a basic piece of mathematical symbolism; for example, $5 + 3$ contains three mathematics tokens: 5, +, and 3. Letters, as used in algebra, and particular mathematical signs are also mathematics tokens, so that x^2 contains the two mathematics tokens, x and ^{2}. Word tokens are ordinary words, as used in English prose. The modified cloze procedure adopted was to delete every fifth token, mathematics tokens being replaced by short lines, and word tokens by long lines, so that a subject who was filling in the blanks could tell whether a word token or a mathematics token was intended to fill the blank space.

To give a British example, Figure 7.4, p. 87, shows a possible cloze passage constructed from *SMP Revised Advanced Mathematics*, Book 1, page 18. A passage from an advanced text has been chosen to show how a wide variety of tokens can be replaced by blanks.

In the course of Kane, Byrne and Hater's work, the relative difficulty of a considerable number of passages taken from American mathematics textbooks became known to them, because comprehension tests and cloze tests had been set on these passages. A study was also undertaken to find which mathematical words and mathematics tokens were well known, and which were relatively unknown, to large numbers of American children in grades 7 and 8 (ages 12 and 13); lists of these well-known and little-known words and tokens were drawn up. It was perhaps surprising that the only mathematics tokens known by 90 per cent of children were +, −, ×, $, ÷, %, ¢, and the numerals.

Kane *et al.* correlated a large number of features of the texts with the reading ease of these texts, and devised a linear combination of the most important of these features which would predict reading ease. Two formulae are given in *Helping Children Read Mathematics*, and the second of these is regarded by the authors as the more reliable. The use of both formulae involves consulting lists of well-known and little-known mathematical words and symbols which are given in *Helping Children Read Mathematics.* These lists are lengthy, and so are not reproduced here. Kane, Byrne and Hater's Formula II is as follows:

$$\text{Predicted Readability} = -0.15A + 0.10B - 0.42C - 0.17D + 35.52$$

where the values of the variables A, B, C, and D are obtained as follows:

A Words not on the Dale list of 3000 common (non-mathematical) words and also not on the list 'Mathematics Words Known to 80 per cent of Children'.
B Number of changes from a word token to a mathematics token and vice versa in the 400-token passage selected.
C Number of different mathematics terms not on the list 'Mathematics Words Known to 80 per cent of Children', plus the number of different mathematics symbols not on the list 'Symbols Known to 90 per cent of Children'.
D Number of question marks in the 400-token passage.

It is recommended that in order to assess the readability of a book, at least ten samples of its text, each consisting of 400 tokens, should be counted, and their mean readability score found. The formula suffers from the same defect as some of the standard readability formulae for OE: it does not give a predicted reading age, but it does provide a readability score which can be used to put different texts in order of difficulty. The formula is so arranged that the larger the readability score, the easier the text is to read; a more difficult text has a lower readability score. The recommended procedure is a tedious one to carry out by hand, as it involves constantly comparing the text with lists of words. For this reason alone it would not seem to be a practical tool for the busy teacher.

The formulae suffer, too, from other problems. They were devised by research with American children, using American curricula in the early 1970s, and they do not seem to be culture-free or curriculum-free. The lists of mathematics words and tokens on which the tests are based were obtained by searching American mathematics texts of the time, written for children in the 7th and 8th grades. The usual curriculum in American schools postpones work in algebra and geometry to grades later than the 8th, so that the 7th and 8th grade curriculum is usually heavily oriented towards arithmetical ideas. Hence, the words *tessellation* and *mapping,* which are likely to be known to large numbers of British children in the 1980s, do not appear at all in the lists of mathematical words. Thus, it seems unlikely that these formulae could be taken over for use in Britain without modification. Their importance lies in the fact that it has been found possible to devise readability formulae which take account of some of the specifically mathematical features of children's mathematics textbooks, as well as those features which occur in texts written only in ordinary English.

(a)

The successive terms of a sequence may be written

$u_1, u_2, u_3, \ldots, u_n, \ldots.$

A sequence may be defined
(i) by a formula for u_n that holds for all natural numbers,
(ii) by an inductive definition where u_1 is given together with an equation connecting u_{k+1} and u_k that holds for all $k \in N$.
(N is the set of natural numbers $\{1, 2, 3, \ldots\}$.)
An *arithmetic progression* is a sequence such that

$$u_1 = a$$
$$u_{k+1} = u_k + d \text{ for all } k \in N,$$

where a and d are any two numbers.

(b)

The successive ______ of a sequence may ______ written

$u_1, u\_\_\_, u_3, \ldots, u_n, \ldots.$

______ sequence may be defined
(i) ______ a formula for u ___ that holds for all ______ numbers,
(ii) by an inductive ______ where u_1 is ______ together with an equation ______ u_{k+1} ______ u_k that holds ______ all $k \in N$.
(___ is the set of ______ numbers $\{1, 2, 3, \ldots\}$.)
______ *arithmetic progression* is a ______ such that

$$u_1 \;\_\_\_\; a$$
$$u_{k+}\_\_\_ = u_k + \_\_\_ \text{ for all } k \in \_\_\_,$$

where a and d ______ any two numbers.

Figure 7.4 (a) The original passage. (b) The cloze passage (every fifth token deleted).

In order to look at the compatibility of Kane, Byrne and Hater's Formula II with present-day British textbooks, we have tried it on the two parallel versions of *Oxford Comprehensive Mathematics.* This series provides a single Book 1 and Book 2, followed by two sets of parallel books – Blue Books 3, 4, 5, which are intended for pupils who are following an O-level course, and Green Books 3, 4, 5, intended for those who are following a CSE course. It was expected that parallel books in the same series, although not by the same authors, might reveal differences in readability corresponding to the different abilities of their intended readers. The Blue and Green versions of Book 3 were chosen for study, because the third year of secondary schooling in England is a stage comparable with grade 8 in the USA.

Five samples were randomly chosen from each version of Book 3, and the following readability scores were obtained by using Kane, Hater and Byrne's Formula II:

Blue Book (O-level): 30.25 30.74 25.73 22.32 35.44
Mean readability score = 28.90
Standard deviation = 5.03
Green Book (CSE): 27.55 25.96 33.56 33.69 28.51
Mean readability score = 29.81
Standard deviation = 3.58

Thus, although the Green Book gave a marginally higher readability score, and so might be marginally easier, the variation between different samples of the same text was much greater than the variation between the mean scores of the two texts. It might therefore be predicted that readers of the CSE text, whose average reading age is lower, would find their text more difficult to read than their O-level colleagues would find theirs.

Both the easiest passage and the most difficult passage were found in the O-level book. The 'easiest' passage was on Inequalities, and contained many changes from word tokens to mathematics tokens. There were many passages such as the following:

'On a number line, draw lines to show: 1. $X = \{x | x < 4\}$'

The 'most difficult' passage was on Isometries, and much of its difficulty was caused by a very large proportion of words which were neither on the Dale list of 3000 familiar English words nor on the list of mathematical words which were well known to 80 per cent of American children. These words included:

transformation	reflection	congruent
rotation	isometry	tangram

Vocabulary such as this is special to a particular mathematics course, and it seems quite likely that pupils who have followed a particular course for a number of years might well not find words such as these as difficult as the method used to devise Formula II assumes them to be. Similarly, it is not clear how easily readable is the constant repetition of formulae such as:

$$X = \{x | x < 4\}$$

once their initial unfamiliarity has been overcome. Kane, Byrne and Hater are aware of the limitations of all readability formula, and they advise that:

> they be used in conjunction with the judgments of teachers, curriculum workers and specialists in mathematical education.

It would seem that more research on the development of tests of readability in mathematics is needed, if such tests are to be of use in this country. If tests of readability in mathematics are not curriculum-free, this would seriously limit their usefulness both for research purposes and to teachers wishing to assess the relative readability of different textbook series which they examine with a view to purchase.

8 The visual appearance of text

Factors which affect visual appearance

The appearance of a page of text depends upon factors which are not usually consciously considered by the teacher or the pupil, except in a purely qualitative way when the visual appeal of the page strikes the reader. However, there are several factors which influence the appearance of a text. It seems likely that the visual appearance of a good page of text will be:

- easy for the reader to find his way about
- pleasing to look at

These qualities can be achieved through a careful choice of:

- the layout of the page
- the type style used in printing
- the use of colour

The teacher considering a textbook for possible adoption has no control over these features, but he needs to be aware that the visual appearance of a book can make a considerable difference to the enthusiasm or otherwise with which pupils approach it.

The layout of the page

An attractive page presents a clean, uncluttered appearance, and is laid out clearly so that the reader will find his way easily around the page. However, some compromises are necessary in the production of a textbook since a lavishly illustrated large-print volume will inevitably be expensive and the school may not be able to afford it. The textbook publisher always has to make compromises between achieving ideal attractiveness of layout and producing a book which will appeal to teachers as 'good value for money'.

We get an idea of the difference which a larger page size and careful attention to layout can make to the attractiveness of the page, by looking at the pages shown in Figures 8.1 and 8.2. These two pages of *Mathematics for Schools,* Level II, Book 4, are corresponding pages from the first (1971) and second (1980) editions. Each page is reduced to 70 per cent of its original dimensions. Not only has the prose reading matter been simplified in the second edition, but the larger page size has allowed the layout to be more carefully planned. Some of the cluttered and unnecessary signals, such as **a]**, have also been removed, and an altogether more attractive page has been produced.

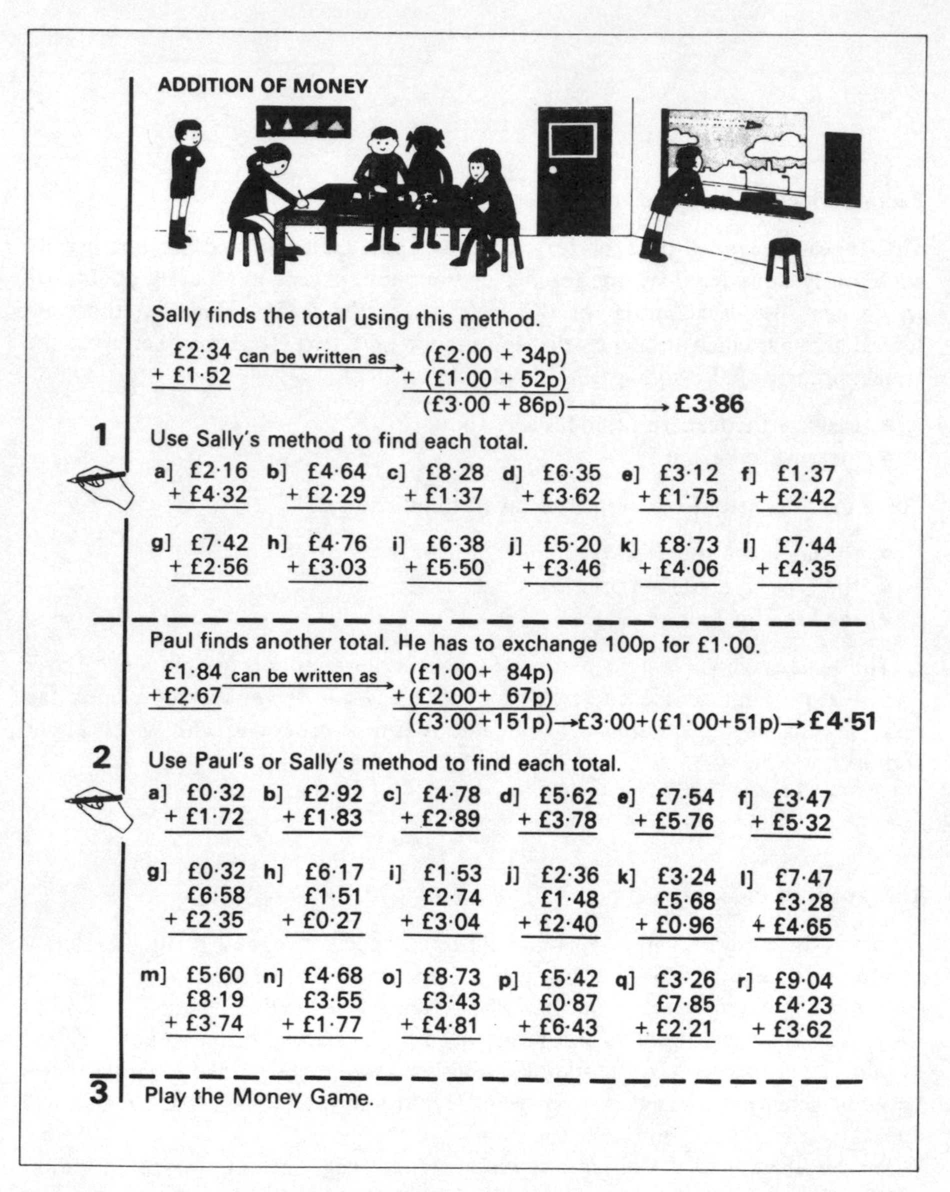

ADDITION OF MONEY

Sally finds the total using this method.

£2·34 + £1·52 can be written as (£2·00 + 34p) + (£1·00 + 52p) = (£3·00 + 86p) → **£3·86**

1 Use Sally's method to find each total.

a] £2·16 + £4·32 b] £4·64 + £2·29 c] £8·28 + £1·37 d] £6·35 + £3·62 e] £3·12 + £1·75 f] £1·37 + £2·42

g] £7·42 + £2·56 h] £4·76 + £3·03 i] £6·38 + £5·50 j] £5·20 + £3·46 k] £8·73 + £4·06 l] £7·44 + £4·35

Paul finds another total. He has to exchange 100p for £1·00.

£1·84 + £2·67 can be written as (£1·00+ 84p) + (£2·00+ 67p) = (£3·00+151p) → £3·00+(£1·00+51p) → **£4·51**

2 Use Paul's or Sally's method to find each total.

a] £0·32 + £1·72 b] £2·92 + £1·83 c] £4·78 + £2·89 d] £5·62 + £3·78 e] £7·54 + £5·76 f] £3·47 + £5·32

g] £0·32, £6·58 + £2·35 h] £6·17, £1·51 + £0·27 i] £1·53, £2·74 + £3·04 j] £2·36, £1·48 + £2·40 k] £3·24, £5·68 + £0·96 l] £7·47, £3·28 + £4·65

m] £5·60, £8·19 + £3·74 n] £4·68, £3·55 + £1·77 o] £8·73, £3·43 + £4·81 p] £5·42, £0·87 + £6·43 q] £3·26, £7·85 + £2·21 r] £9·04, £4·23 + £3·62

3 Play the Money Game.

Figure 8.1 *Mathematics for Schools*, Level II, Book 4, page 6 (first edition).

A book in large print with good illustrations has some disadvantages, as the experience of the School Mathematics Project shows. The first SMP texts for the

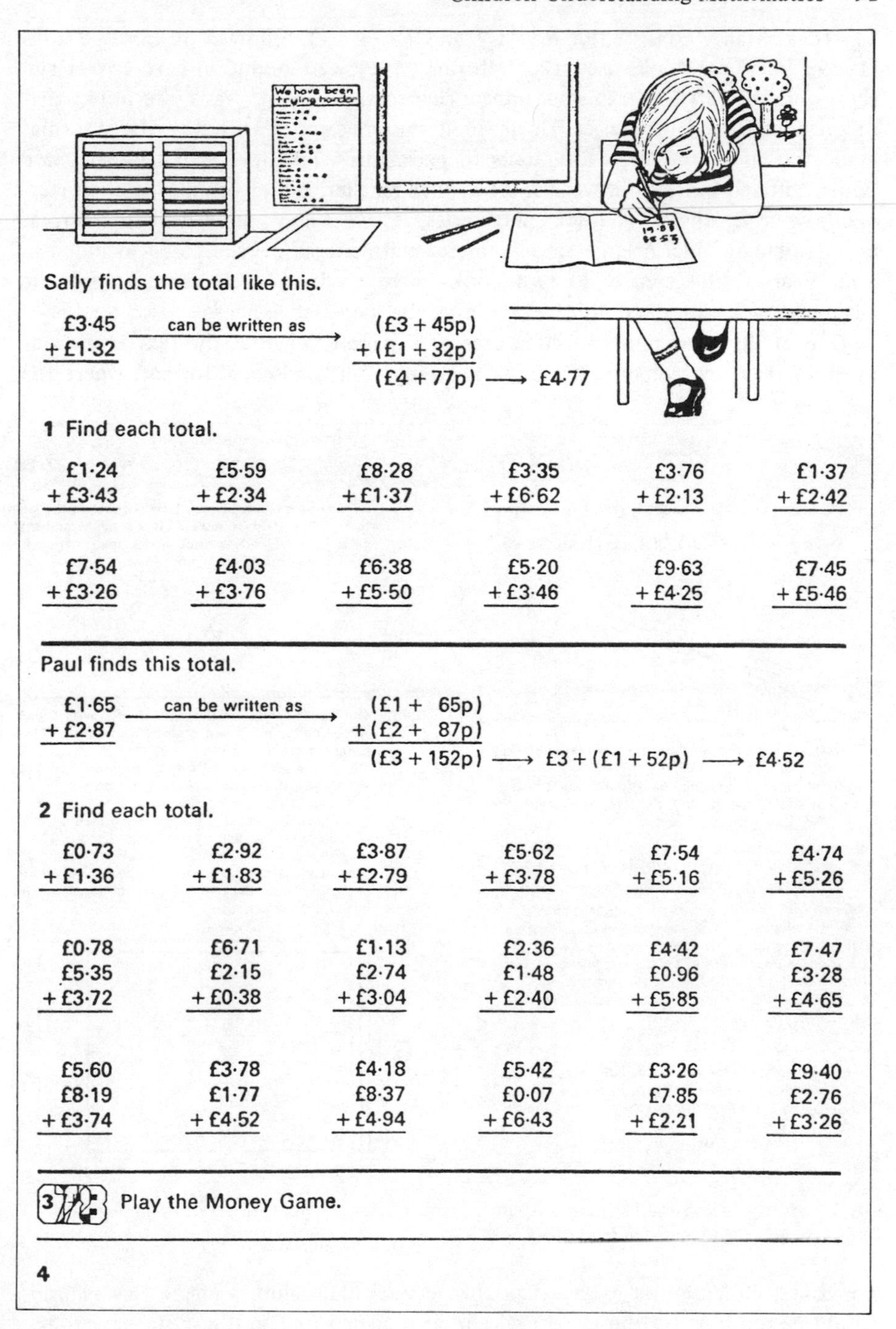

We have been trying harder

Sally finds the total like this.

£3·45 + £1·32 —can be written as→ (£3 + 45p) + (£1 + 32p) = (£4 + 77p) ⟶ £4·77

1 Find each total.

£1·24	£5·59	£8·28	£3·35	£3·76	£1·37
+ £3·43	+ £2·34	+ £1·37	+ £6·62	+ £2·13	+ £2·42

£7·54	£4·03	£6·38	£5·20	£9·63	£7·45
+ £3·26	+ £3·76	+ £5·50	+ £3·46	+ £4·25	+ £5·46

Paul finds this total.

£1·65 + £2·87 —can be written as→ (£1 + 65p) + (£2 + 87p) = (£3 + 152p) ⟶ £3 + (£1 + 52p) ⟶ £4·52

2 Find each total.

£0·73	£2·92	£3·87	£5·62	£7·54	£4·74
+ £1·36	+ £1·83	+ £2·79	+ £3·78	+ £5·16	+ £5·26

£0·78	£6·71	£1·13	£2·36	£4·42	£7·47
£5·35	£2·15	£2·74	£1·48	£0·96	£3·28
+ £3·72	+ £0·38	+ £3·04	+ £2·40	+ £5·85	+ £4·65

£5·60	£3·78	£4·18	£5·42	£3·26	£9·40
£8·19	£1·77	£8·37	£0·07	£7·85	£2·76
+ £3·74	+ £4·52	+ £4·94	+ £6·43	+ £2·21	+ £3·26

3 Play the Money Game.

4

Figure 8.2 *Mathematics for Schools,* Level II, Book 4, page 4 (second edition).

11–16 age range consisted of *Books T* and *T4* (1964), followed by *Books 1* to *5* (1965). All these books used a large format, they were bound in hard covers and they were very attractive in appearance; their general layout was quite unlike that of previous O-level text series. However, it soon became clear that the large format made the books too large for pupils to pack easily into their satchels; the SMP books suffered more from wear and tear than smaller, lighter volumes, and replacements were expensive. The next SMP series, *Books A* to *H*, used a smaller format and flexible binding. A large amount of textual material needed to be included in each year of the course, so two books were used for each year: against the advantage of greater durability had to be set the increased expense of two books.

One of the most noticeable features of a page is whether the text is laid out in more than one column. If a book is arranged in 'landscape' format, where the

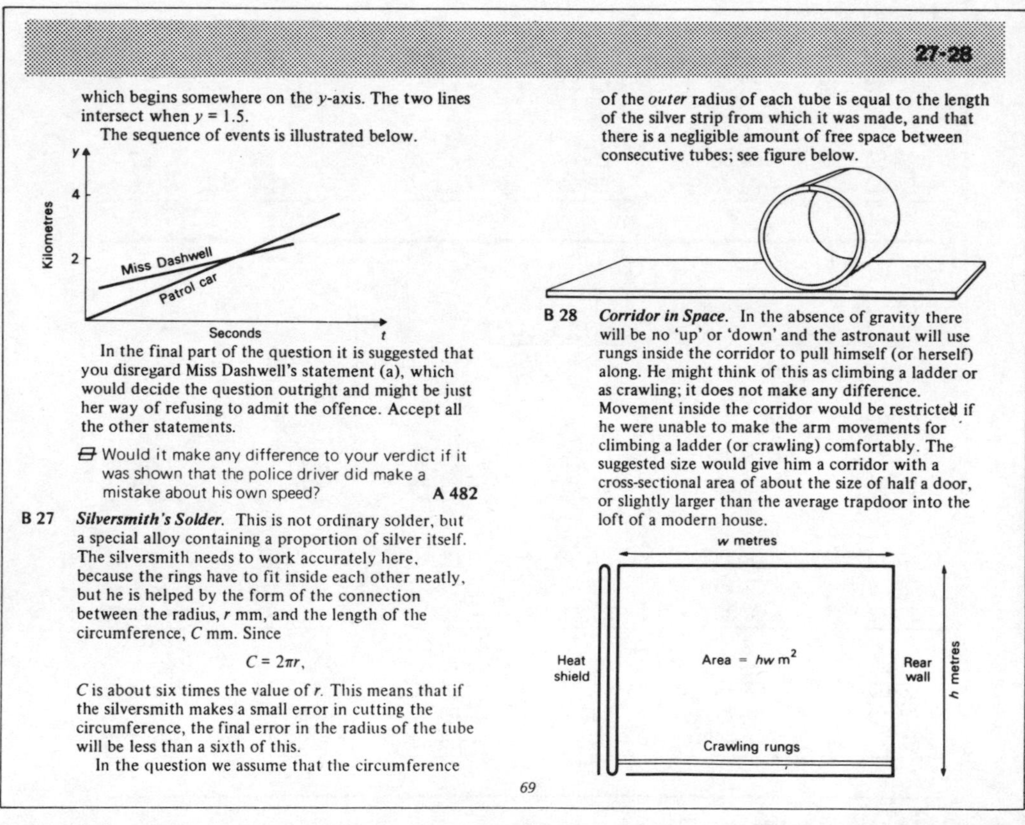

27-28

which begins somewhere on the y-axis. The two lines intersect when $y = 1.5$.

The sequence of events is illustrated below.

In the final part of the question it is suggested that you disregard Miss Dashwell's statement (a), which would decide the question outright and might be just her way of refusing to admit the offence. Accept all the other statements.

Would it make any difference to your verdict if it was shown that the police driver did make a mistake about his own speed? **A 482**

B 27 ***Silversmith's Solder.*** This is not ordinary solder, but a special alloy containing a proportion of silver itself. The silversmith needs to work accurately here, because the rings have to fit inside each other neatly, but he is helped by the form of the connection between the radius, r mm, and the length of the circumference, C mm. Since

$$C = 2\pi r,$$

C is about six times the value of r. This means that if the silversmith makes a small error in cutting the circumference, the final error in the radius of the tube will be less than a sixth of this.

In the question we assume that the circumference of the *outer* radius of each tube is equal to the length of the silver strip from which it was made, and that there is a negligible amount of free space between consecutive tubes; see figure below.

B 28 ***Corridor in Space.*** In the absence of gravity there will be no 'up' or 'down' and the astronaut will use rungs inside the corridor to pull himself (or herself) along. He might think of this as climbing a ladder or as crawling; it does not make any difference. Movement inside the corridor would be restricted if he were unable to make the arm movements for climbing a ladder (or crawling) comfortably. The suggested size would give him a corridor with a cross-sectional area of about the size of half a door, or slightly larger than the average trapdoor into the loft of a modern house.

69

Figure 8.3 Schools Council Sixth Form Mathematics Project, *Polynomial Models* (1978), page 69 (reduced to 65% of original size).

horizontal dimension is longer than the vertical dimension, a single line of print would be too long for the eye to take in at a glance, and so it is customary to use more than one column. Figure 8.3 shows an example, taken from the *Mathematics Applicable* series, and intended for sixth-formers. Another example is shown in

Figure 8.4. This is a page from *Maths Adventure, Book 3,* for children of about 9 to 10 years old; the division of the page into three columns produces a layout in which many short questions can be fitted on to the page. The layout of this page, however, often confuses pupils who expect to read down a page to the bottom of a column before moving to the next column. Indeed, questions 1 and 2 encourage the child to read down the column, and he may easily not notice that the next question in that column is in fact number 5. In this example, the signals provided by the question numbers are most important in steering the reader around the page.

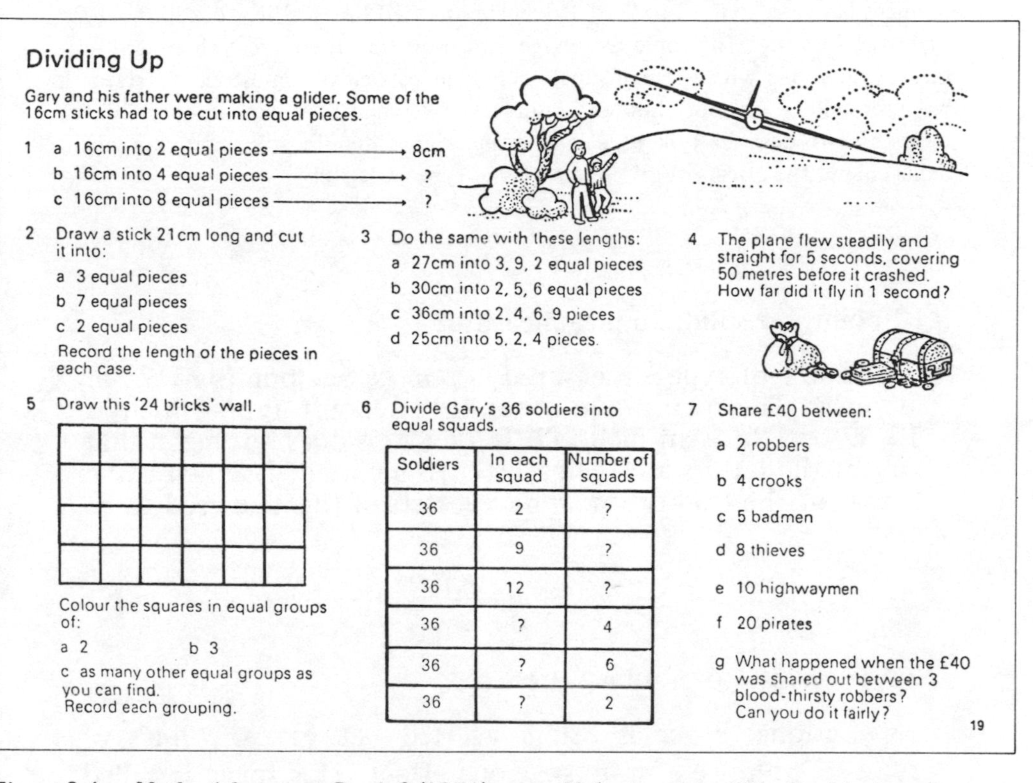

Dividing Up

Gary and his father were making a glider. Some of the 16cm sticks had to be cut into equal pieces.

1 a 16cm into 2 equal pieces → 8cm
b 16cm into 4 equal pieces → ?
c 16cm into 8 equal pieces → ?

2 Draw a stick 21cm long and cut it into:
a 3 equal pieces
b 7 equal pieces
c 2 equal pieces
Record the length of the pieces in each case.

3 Do the same with these lengths:
a 27cm into 3, 9, 2 equal pieces
b 30cm into 2, 5, 6 equal pieces
c 36cm into 2, 4, 6, 9 pieces
d 25cm into 5, 2, 4 pieces.

4 The plane flew steadily and straight for 5 seconds, covering 50 metres before it crashed. How far did it fly in 1 second?

5 Draw this '24 bricks' wall.

Colour the squares in equal groups of:
a 2 b 3
c as many other equal groups as you can find.
Record each grouping.

6 Divide Gary's 36 soldiers into equal squads.

Soldiers	In each squad	Number of squads
36	2	?
36	9	?
36	12	?
36	?	4
36	?	6
36	?	2

7 Share £40 between:
a 2 robbers
b 4 crooks
c 5 badmen
d 8 thieves
e 10 highwaymen
f 20 pirates
g What happened when the £40 was shared out between 3 blood-thirsty robbers? Can you do it fairly?

19

Figure 8.4 *Maths Adventure, Book 3* (1971), page 19 (reduced to 60% of original size).

The way the diagrams and illustrations are arranged in relation to the text is also a very noticeable feature of layout. There has been some theoretical discussion of how the placing of diagrams affects readability – is it better to place diagrams to the right or the left, or above or below, the related text? Research done by the Typography Unit of the University of Reading (Smith and Watkins, 1972) indicated that there was little or no significant difference in the comprehension of a passage when the relative positions of the text and diagrams were varied. The presence of the illustrations themselves did, however, cause a significant improvement in learning. It is also important that the diagram should be placed close to the text to which it refers (see p. 53).

The type styles used in printing

In this section we describe some features of the different styles of type which are found in mathematics textbooks, and look at their effect on the reader. The effect on the ease of reading caused by the style and layout of the type is usually termed *legibility*; the term *readability,* which we have discussed previously, relates to the content of the text rather than to the type.

The authority on the legibility of OE text is Tinker. He spent over thirty years studying legibility, using many thousands of subjects in his comparisons of differences between styles. In 1965, Tinker summarised some of his work in *Bases for Effective Reading.* The topic is so large, however, that there are still some fundamental questions which remain unanswered; in particular, almost all the research has used adults and not children as subjects. We shall now explain some of the terms used to describe type used in printing, and summarise some commonly held beliefs about the effects of different styles of type on legibility.

12 point, set solid, 26 pica measure:

The size of type is measured in *points*; one point is 1/72 of an inch. This paragraph is set in 12 point type which is 12/72 or 1/6 of an inch tall. In fact, this does not mean that any of the letters are themselves 1/6 inch tall; 1/6 inch is the height of the lead type in which letters of this size used to be cast.

12/14 point, 26 pica measure:

Additional space is often inserted between the lines of type to produce a more spacious effect. This space is called *leading*. Thus, 12 point type may be 'set solid' (no space) or with 1 point leading (an extra 1/72 inch between lines), $1\frac{1}{2}$ point leading, and so on. A text set in 12 point type with 2 point leading (as this paragraph is set) is often called 12/14 or '12 on 14'. The length of a line of text is often measured in *picas*; a pica is 1/6 inch or 12 points.

For average secondary pupils, 11 or 12 point type is widely held to be the best size of type, and $1\frac{1}{2}$ or 2 points of leading significantly improve the ease of reading.

It is not a good idea for the publisher to try to cram more text on to the page by using smaller point size or less leading; this will cause loss of concentration and make the text difficult for pupils to read.

The best line length depends on the type size and leading. Lines which are too short cause unnecessary interruptions to eye-movement, while lines which are too long tend to cause the eye to lose its place. Between 20 and 30 picas of 11 or 12 point type seem to be suitable in text for adults.

The letters themselves have different shapes in different typefaces. Some typefaces have *serifs,* some are *sans serif.*

Times Roman has serifs,

Univers is a sans serif type face.

Some typefaces have a comparatively small x-height (the height of a letter x).

A line of 11 point Bembo has a small x-height

and so it appears smaller than

a line of 11 point Univers, which has the same overall size, but a larger x-height.

The typeface used in this book is 10 point IBM Press Roman.

Because of the smaller x-height of Bembo the ascenders and descenders of letters such as b, h and p seem more pronounced in Bembo.

There has been a good deal of argument over the merits of serif versus sans serif typefaces. There seems to be no convincing evidence that serifs either increase or decrease legibility. Variables such as the amount of enclosed space in letters such as e and d, the relative height of ascenders and descenders and the thickness of the letters have as much or more influence on legibility than serifs. However, text for primary pupils is often set in sans serif typefaces, because sans serif type is more like written script; secondary texts often use typefaces with serifs, which seem to have a more adult appearance. Figure 8.4 shows an extract from a primary text which uses a sans serif typeface, while the extract in Figure 8.3 is from a secondary text which has serifs. We may notice the way in which a sans serif typeface is used for contrast on this page. Sans serif type has become more popular in general printing, and has shed its previous 'childish' image; in recent years, several secondary mathematics textbooks have been printed in sans serif typefaces.

Some typefaces use confusingly similar characters. It is important in the printing of mathematics to ensure that the digit 1 is not easily confused with the letters I and l, as they can be in the typeface used for this book, and that the multiplication sign × and the letter x do not look alike. The two extracts from *Mathsworks*, shown in Figure 8.5, are intended for secondary school pupils of below-average attainment. They show how alike the multiplication sign and the letter x look in the typeface used. The question shown in Figure 8.6, from a sixth-form text, Porter's *Further Elementary Analysis*, shows how easily the figure 1 in the formulae can be confused with the letter l.

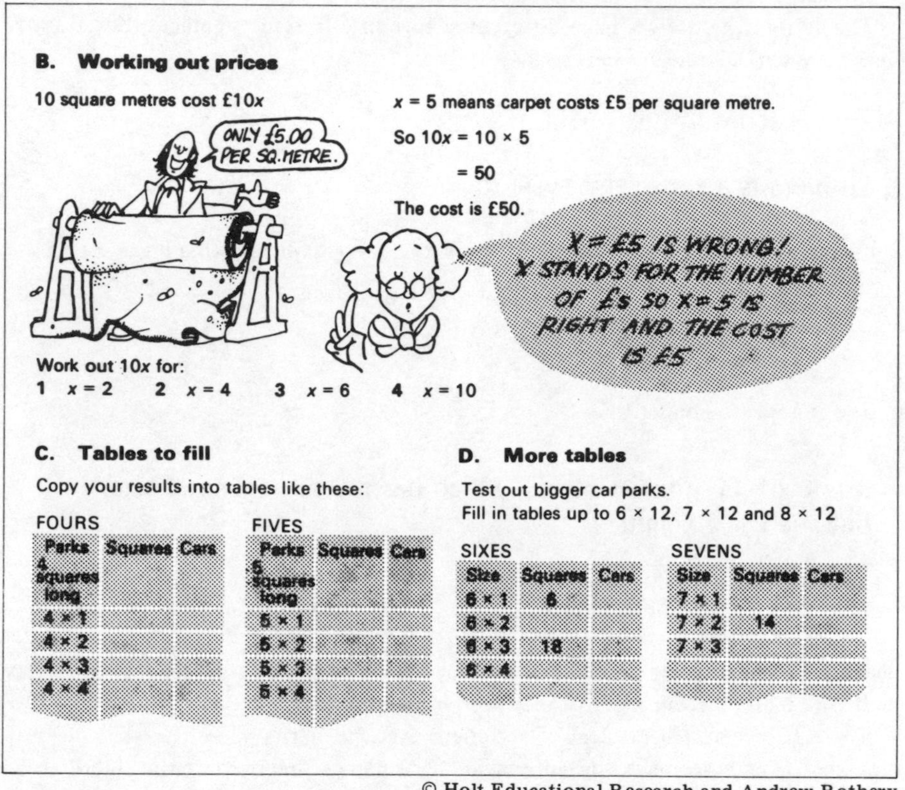

B. Working out prices

10 square metres cost £10*x*

ONLY £5.00 PER SQ. METRE.

x = 5 means carpet costs £5 per square metre.

So 10*x* = 10 × 5

= 50

The cost is £50.

X = £5 IS WRONG! X STANDS FOR THE NUMBER OF £s SO X = 5 IS RIGHT AND THE COST IS £5

Work out 10*x* for:

1 *x* = 2 2 *x* = 4 3 *x* = 6 4 *x* = 10

C. Tables to fill

Copy your results into tables like these:

FOURS

Parks 4 squares long	Squares	Cars
4 × 1		
4 × 2		
4 × 3		
4 × 4		

FIVES

Parks 5 squares long	Squares	Cars
5 × 1		
5 × 2		
5 × 3		
5 × 4		

D. More tables

Test out bigger car parks.

Fill in tables up to 6 × 12, 7 × 12 and 8 × 12

SIXES

Size	Squares	Cars
6 × 1	6	
6 × 2		
6 × 3	18	
6 × 4		

SEVENS

Size	Squares	Cars
7 × 1		
7 × 2	14	
7 × 3		

Figure 8.5 Extracts from *Mathsworks, Book One* (1979).

26. Prove that any point whose x and y coordinates satisfy the equations

$$\frac{1-x/a}{t^2} = \frac{1+x/a}{1} = \frac{y/b}{t},$$

where t is a parameter, lies on the ellipse $x^2/a^2 + y^2/b^2 = 1$.

Show also that the coordinates of the point of intersection of the tangents to the ellipse at the points $t = t_1, t = t_2$ satisfy the equations

$$\frac{1-\dfrac{x}{a}}{t_1 t_2} = \frac{1+x/a}{1} = \frac{y/b}{\frac{1}{2}(t_1+t_2)}.$$

Figure 8.6 From Porter, *Further Elementary Analysis* (1970), p. 197.

The example of Figure 8.6 also illustrates another of the problems of printing mathematics: the difficulty of printing expressions which contain division. In his own written work, the student will write the equation of an ellipse as:
$\frac{x^2}{a^2} + \frac{y^2}{b^2} = 1$, but the printer has managed to compress the equation onto a single line by using the solidus / for division, writing:

$$x^2/a^2 + y^2/b^2 = 1.$$

Sixth-form students are unlikely to be baffled by the use of the solidus, although they need to translate it into a horizontal line as they begin their own written work. However, in this particular case, the student may be baffled by the inconsistent use which the printer has made of the solidus in the formula

$$\frac{1 - \frac{x}{a}}{t_1 t_2} = \frac{1 + x/a}{1} = \frac{y/b}{\frac{1}{2}(t_1 + t_2)}$$

It is advisable to avoid the use of the solidus as a division or fraction sign altogether in text for younger pupils. The less able of the secondary pupils who use Holt's *Maths 3* (Figure 8.7) may well not realise that 6 ½ is equal to $6\frac{5}{10}$.

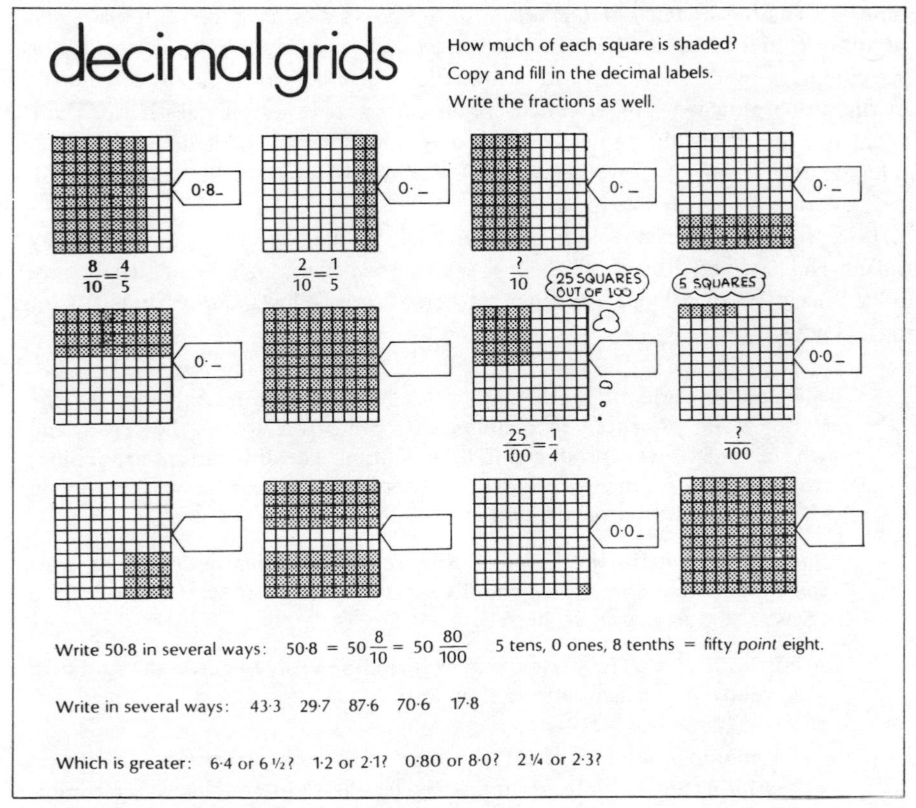

Figure 8.7 From *Maths 3* (1974), page 38.

There is quite a lot of evidence that text set entirely in capital letters is much less legible than text set in lower case. For this reason, road signs are now printed in lower case rather than upper case. For emphasis, a **bold typeface** is easier to read than an *italic typeface.* Naturally, we do not expect to find a whole book set in either of these, but the differences can be borne in mind when inspecting a text.

We can appreciate the problems which Tinker and others face in trying to isolate the effects of particular variables on legibility, and it is difficult, therefore, to provide guidance for the teacher. Certainly, a good text should not be rejected purely on the grounds of a poor choice of typeface. However, when inspecting a pupils' text for possible use, it may be helpful to bear in mind the effects which different styles of printing are believed to have on legibility.

The use of colour

One of the least quantifiable factors which contributes to the appearance of a page is the use of colour. The publisher decides, partly on economic grounds, the number of colours to be used in a book, each colour requiring an additional print stage. Many books are printed in black on white (or rarely, some other colour on white). Some, such as the Fletcher *Mathematics for Schools* series, are printed in black and one other colour on white. This is called two-colour printing – black counts as the first colour. A more costly method is full colour printing, when almost any hue of colour can be provided. The effect of colour on a page is not only aesthetic. There is evidence that the addition of colour to diagrams can lead to significantly better understanding (Smith and Watkins, 1972). Children prefer bright colours to dull ones, and they especially like red and blue.

However, an unnatural colour scheme, such as blue grass and red sky, may confuse and distract. King (1978), in a sociological study of infants' classrooms, noticed the effect of colour in Fletcher mathematics as follows, quoted in part from his field notes:

> The Fletcher world of mathematics has a superficial resemblance to the 'real-life' world presented to children in some of the stories they read, and which they may have reproduced in their writing. The illustrations are similar, showing children, animals, food, toys, aeroplanes, and cards . . . However, in the world of mathematics conventional reality may be suspended.
>
> > Sharon does maths work book. She follows the instructions to colour some dolls blue and some red. 'I've never seen people with blue or red faces', she says, mainly to herself.
> >
> > A boy sorts sets using plastic shapes including yellow Scottie dogs, orange pigs, yellow horses, and purple elephants.

Colour is sometimes used to print the text of a book, sometimes on tinted paper. Black printing on a white ground is by far the most legible colour combination, because there is the greatest possible contrast between the printing and the background. In general, a bold colour on a very faintly coloured background will be slightly less legible. Coloured print may produce interest, as a welcome change from the usual black on white, or it may be used to emphasise important passages. However, if the passages to be emphasised are printed in a faint colour such as orange, the emphasis may not succeed; some examples of this can be found in the two-colour printing of the Fletcher series. A strongly coloured background also makes the text very difficult to read.

In general, a good use of colour does help children to read the page easily, and makes it more attractive. This effect is more pronounced with younger children; less able children in particular find good colour in reading material a great help.

Comparison of different layouts

We have tried informally to gauge the effect on pupils of using different layouts in a page of mathematics text. A page on Hire Purchase was prepared for use and two different layouts were designed, one handwritten and using an informal presentation, and the other typed and containing more formal illustrations. The two layouts are shown in Figures 8.8 and 8.9 (pp. 100, 101). They were used by about 500 fourth-year secondary pupils (representing a wide range of ability) in two different mixed comprehensive schools. The pupils were shown the two pages and questioned about their preference, if any. They were then asked to work through the page which they preferred. When they had done this, they were asked to write on the back of the page any comments they wanted to make. The pupils were told that they were helping to decide which type of layout to use in a book. This seemed to ensure that they looked upon the task with some interest rather than regarding it as a test.

Of the 500 pupils, 71 per cent preferred the typewritten version (76 per cent of boys and 65 per cent of girls). It was noticeable that pupils of higher ability showed a definite preference for the typescript; over 90 per cent of the O-level and top CSE sets chose this version. Some of the reasons given were:

It looked neater.
It looked easier to follow.
It looked more like an ordinary maths book.
It looked as if you could follow the worked example easier.

Among the comments from those who chose the handwritten version were:

It looked more like real.
It looked more interesting and less like maths.
You could tell which price label went with what. (This was presumably because price labels were actually attached to items on the handwritten sheet.)

It was noticed that low-ability pupils who chose the handwritten sheet completed the work better than pupils of equal ability who chose the typescript. However, many pupils who chose the handwritten sheet thought that there was a fraction in the second line of the worked example:

$$\frac{£520}{£620}$$

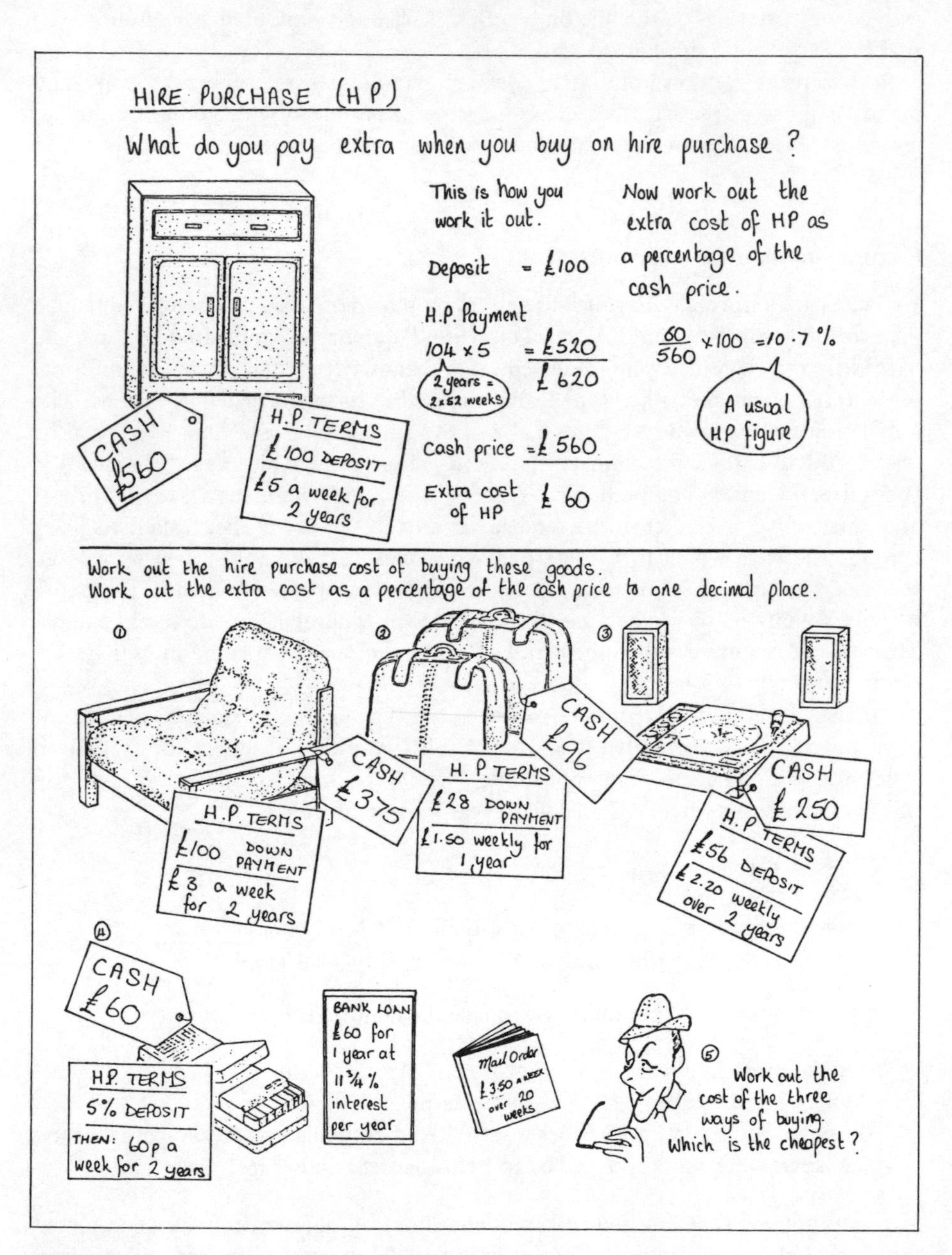

Figure 8.8 The handwritten version.

Another fault in the design is that the layout of the bottom of the typewritten page seems to indicate that the cartoon head belongs to Question 5, whereas he was intended to refer to Question 4 – in fact, Question 4 is hardly a question at all! Several pupils commented that the page was difficult to follow: 'You had to hunt

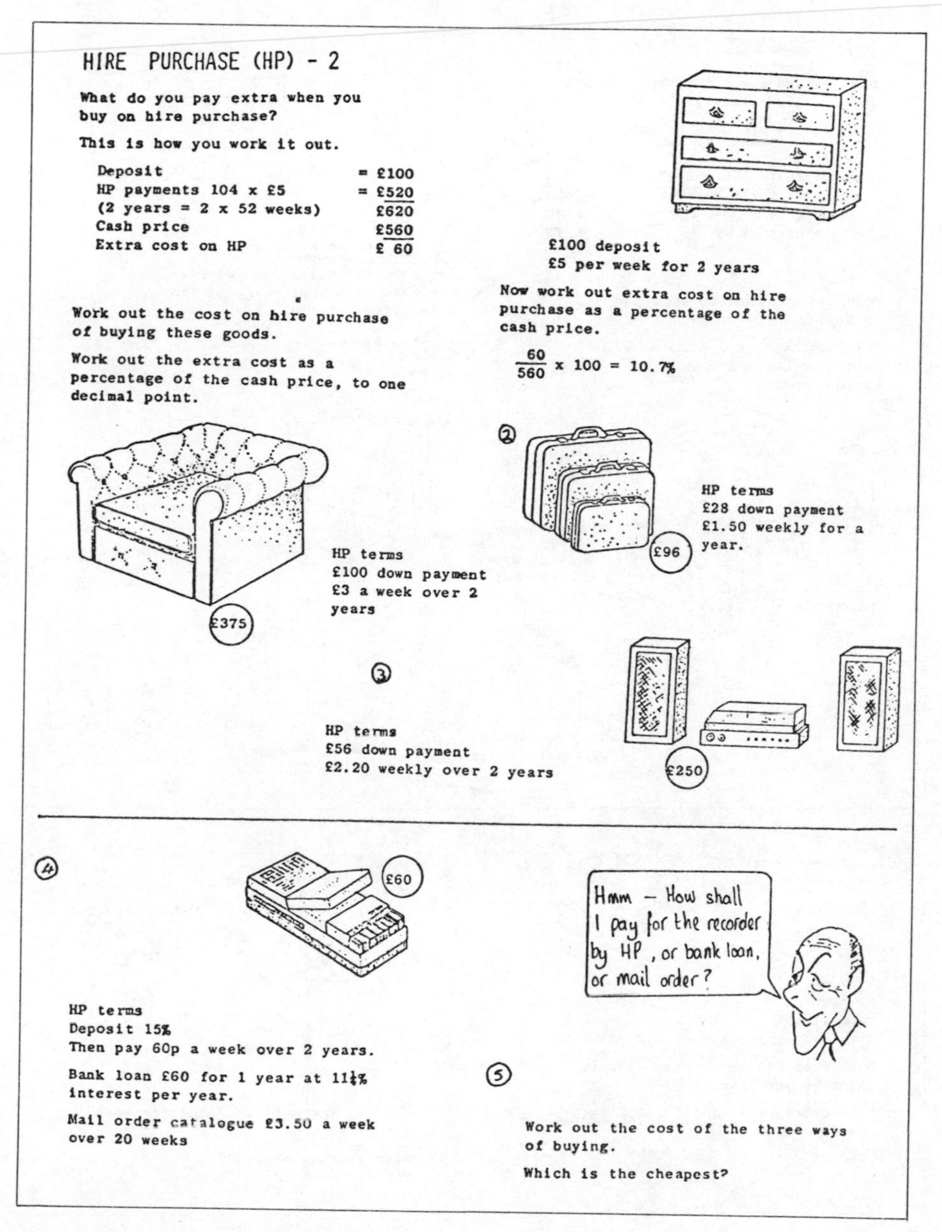

HIRE PURCHASE (HP) - 2

What do you pay extra when you buy on hire purchase?

This is how you work it out.

Deposit	= £100
HP payments 104 x £5	= £520
(2 years = 2 x 52 weeks)	£620
Cash price	£560
Extra cost on HP	£ 60

£100 deposit
£5 per week for 2 years

Now work out extra cost on hire purchase as a percentage of the cash price.

$$\frac{60}{560} \times 100 = 10.7\%$$

Work out the cost on hire purchase of buying these goods.

Work out the extra cost as a percentage of the cash price, to one decimal point.

£375

HP terms
£100 down payment
£3 a week over 2 years

2 £96

HP terms
£28 down payment
£1.50 weekly for a year.

3 £250

HP terms
£56 down payment
£2.20 weekly over 2 years

4 £60

HP terms
Deposit 15%
Then pay 60p a week over 2 years.

Bank loan £60 for 1 year at 11¼% interest per year.

Mail order catalogue £3.50 a week over 20 weeks

5

Work out the cost of the three ways of buying.

Which is the cheapest?

Figure 8.9 The typed version.

around the page to find where to go next.' Figure 8.10, a possible path around the typescript page, demonstrates that the layout is indeed very confusing. All the teachers whose classes were involved commented that in both versions there was

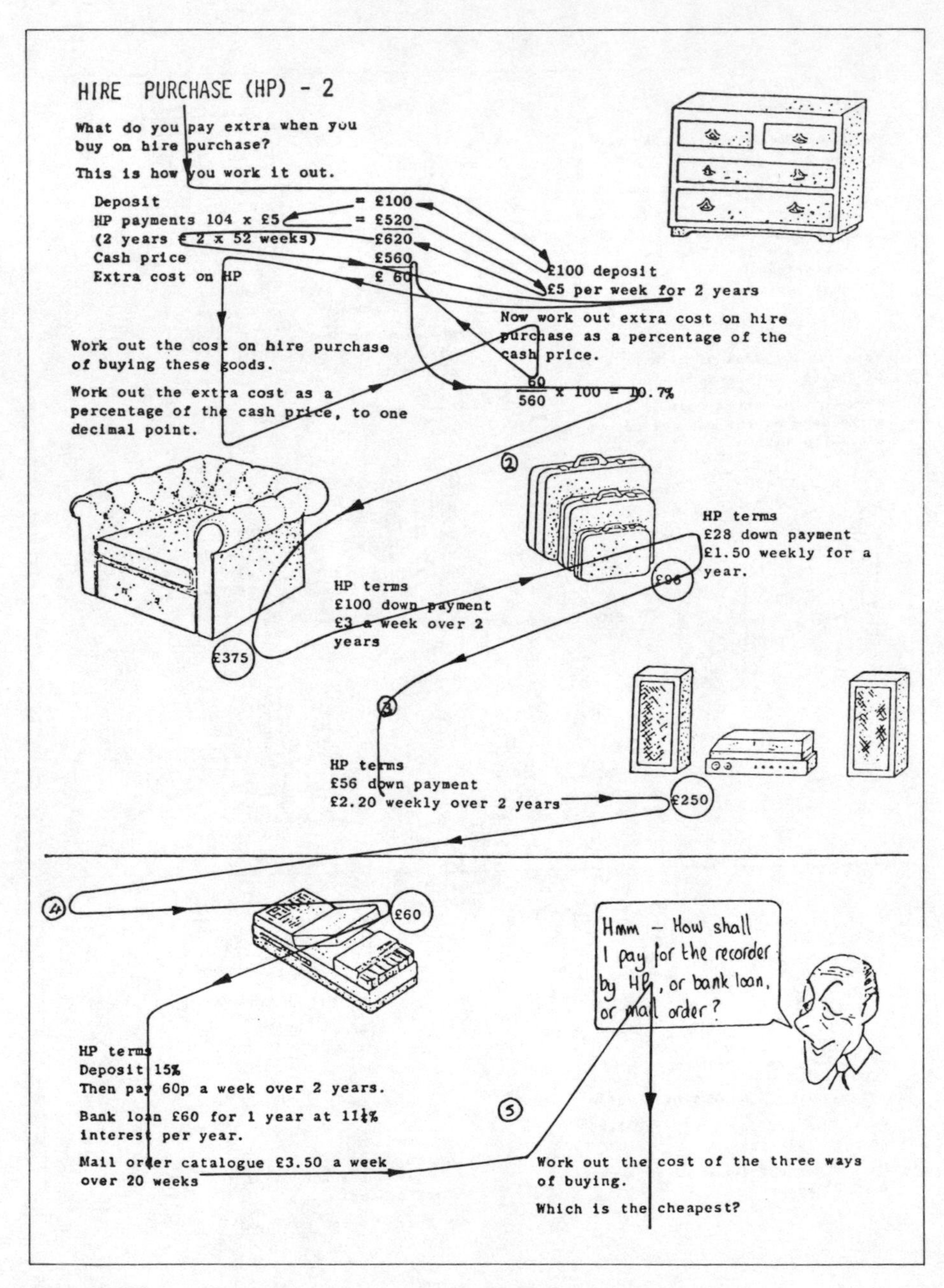

Figure 8.10 A possible path around the page of the typed version.

too much on the page; both layouts clearly need a good deal of improvement. At the same time, the reaction of the pupils to them does seem to indicate that pupils of differing abilities respond differently to the layout of a page. More research in this area would be of assistance both to teachers and to textbook writers.

9 Ideas into practice

Introduction

As we have seen in the previous chapters, there are many factors to be considered in the writing of mathematics text for children, whether the writer is a professional author or an individual teacher writing materials for use in class.

It is most important to ensure that the flow of meaning in the text is clear, and to consider how the graphic material complements and fits in with the verbal material. A careful balancing of graphic and verbal material may sometimes seem wasteful of space on the page, but work with children has convinced us that the visual arrangement of the page is vital.

We shall try to illustrate these points, and points made in the previous chapters, by considering some written material in detail. In the first part of this chapter we analyse a piece of text which was discussed earlier in the book; we then give an account of some informal investigations of alternative ways of presenting the same material which we have undertaken with children.

Analysis of a page of text

The page shown in Figure 9.1 is intended for pupils of below average ability in the fourth year of the secondary school. It contains a great many of the problems of readability which were identified earlier.

Some points which may cause pupils difficulty are the following:

(i) The passage of exposition at the top of the page is badly laid out: the eye has to follow a very difficult path, as shown in Figure 9.2, and there is no guidance on where to look next.

(ii) The layout of the top left-hand box does not make for an easy flow of meaning. Figure 9.3 shows how difficult it is to perceive that five numbers are to be summed to obtain a total of 80; the break in the middle of the sum, and the placing of 16 under =, do not help the eye to see as a group the five numbers that are to be summed.

(iii) The author no doubt intended his language to be informal, but the effect is more strange than informal – 'This is how to work it out by numbers.' What will 'by numbers' mean to the reader?

(iv) The relevance of the top right-hand box is unclear: unless the reader already knows what an average is, this box will mean nothing to him.

(v) The exposition is not clearly separated from the exercises; more space is needed, together with a clearer signal following the passage of exposition, to show where exercises start.

(vi) The meaning of the flag which is intended to encourage the pupil to compare Question 2 with Question 1 is not at all clear: 'TAKE AWAY 10 FROM Q1' may not convey the intended meaning; the pupil may not even divine that 'Q1' means 'Question 1' rather than some strange (and unknown!) mathematical formula.
(vii) In the diagrams of questions 1 and 2, there is some attempt to draw the vertical lines roughly to scale; this attempt is abandoned in questions 3 and 4.
(viii) It is unfortunate that the flag in question 4, 'ADD 100 TO Q1', should read 'ADD 100 TO Q2'.
(ix) The exposition at the top of the page, although referring to lengths, contains no units of length. The note in question 5, 'REMEMBER THE UNITS', could equally well be applied to the exposition! Even if pupils do remember the units, introducing hours and minutes in question 8 is likely to pose problems for low attaining pupils.

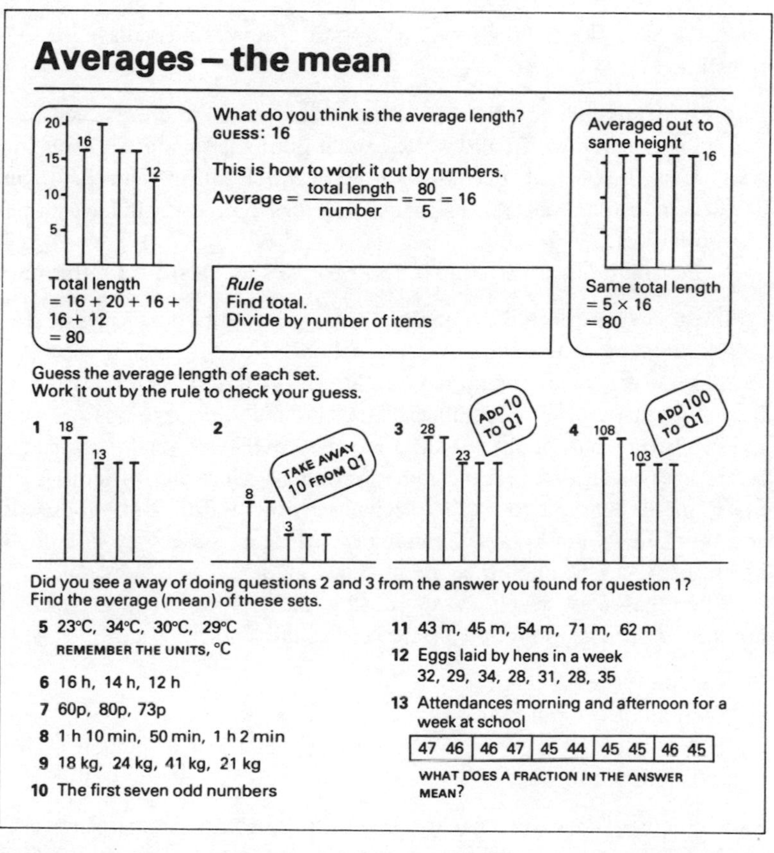

Averages – the mean

What do you think is the average length?
GUESS: 16

This is how to work it out by numbers.
Average = $\frac{\text{total length}}{\text{number}} = \frac{80}{5} = 16$

Total length
= 16 + 20 + 16 + 16 + 12
= 80

Rule
Find total.
Divide by number of items

Averaged out to same height

Same total length
= 5 × 16
= 80

Guess the average length of each set.
Work it out by the rule to check your guess.

1 2 3 4

TAKE AWAY 10 FROM Q1

ADD 10 TO Q1

ADD 100 TO Q1

Did you see a way of doing questions 2 and 3 from the answer you found for question 1?
Find the average (mean) of these sets.

5 23°C, 34°C, 30°C, 29°C
REMEMBER THE UNITS, °C
6 16 h, 14 h, 12 h
7 60p, 80p, 73p
8 1 h 10 min, 50 min, 1 h 2 min
9 18 kg, 24 kg, 41 kg, 21 kg
10 The first seven odd numbers
11 43 m, 45 m, 54 m, 71 m, 62 m
12 Eggs laid by hens in a week
32, 29, 34, 28, 31, 28, 35
13 Attendances morning and afternoon for a week at school

47 46	46 47	45 44	45 45	46 45

WHAT DOES A FRACTION IN THE ANSWER MEAN?

Figure 9.1

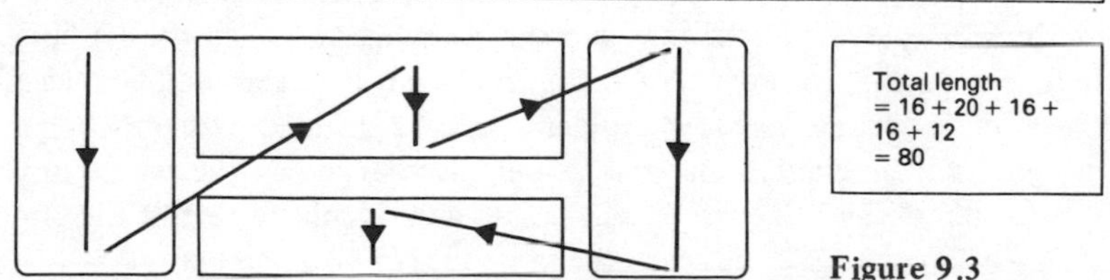

Figure 9.2

Figure 9.3

(x) In question 13 the last two lines:

> WHAT DOES A FRACTION IN THE ANSWER
> MEAN?

are an example of a poor break between lines. This break does not help the pupil's eye as he tries to extract meaning from the page. This is also an example of a question to which no satisfactory answer can be given.

We can see that an analysis of this type can highlight potential problems for the pupil in a page of text, and can suggest revisions that might improve its readability.

Experiments with alternative versions of text passages

Discussions in the writing group, and with other teachers, have led to the following questions:

- Is it possible to write mathematics text for children which has significantly fewer reading difficulties than much published material seems to contain?
- If so, is the text inevitably stilted or much longer than the original, or does it skirt around all the mathematical ideas, or does it have other undesirable features?
- Do children actually find it easier to read and learn from this rewritten text?

In order to explore these questions, we selected two passages of text, one from a well-known series of mathematics textbooks for secondary pupils, and the other from a similar series for primary children; both had been reported by teachers as having given pupils considerable difficulties. Each passage was rewritten, and it was hypothesised that pupils would find the rewritten versions easier to read. The rewritten passages were presented to pupils to see what difficulties emerged, while other pupils worked from the original version. Before the pupils worked through the text, they were briefly reminded of previous work they had done in that area of mathematics, but no other teaching was given.

When passages which cause reading difficulties are discussed with teachers, comments such as the following are very usual:

> 'I wouldn't teach that . . .'
> 'I wouldn't teach it that way . . .'
> 'If I taught it that way, I wouldn't expect them to be able to learn from this text . . .'

Such comments make it clear that it is difficult to discriminate between the teaching methods employed in the text and the language in which the teaching is clothed. Before rewriting, in order to avoid a confusion between mathematical problems in the exposition and language problems, it was decided that only language and the presentation of graphic material would be changed in the rewritten passages; the exposition of the ideas was kept as close as possible to that of the original, so that each version attempted to teach the same ideas in the same way.

Alternative versions of a passage from a secondary text

The passage chosen for rewriting was taken from *SMP, Book B* (1971), Chapter 5, 'Comparison of Fractions'. Because of a time limitation, only the first section of the chapter, pages 51–53, was rewritten. In the end, four versions were produced, and were presented to pupils in similar handwritten format. The four versions are shown in Figures 9.4, 9.5, 9.6, and 9.7. They had the following characteristics:

Version A. This is the original text.
Version B. This was the first attempt at rewriting, and was produced, after much discussion, by the writing group working in committee.
Version C. This was produced by one member of the group, who thought Version B still contained undesirable language features, such as 'red herrings', found in the original text.
Version D. This was produced by the same member of the group, in an attempt to incorporate some of the redundancy which mathematical text usually lacks. This version also differs from the original in that a new emphasis on the order in which coordinates are shown on a graph is introduced.

Approximately 400 first-year secondary school pupils (aged 11–12) were involved in assessing the different versions of the text; they consisted of the complete first-year groups in two mixed comprehensive schools in different parts of the country. All the pupils were given a preliminary reminder about the idea of a fraction, but care was taken not to tell them how to read the materials. The four versions were distributed randomly, and pupils were encouraged to ask questions and to seek clarification from members of staff rather than from their neighbours. They were also asked to write down their comments on the text. Following the initial 50-minute lesson which the pupils worked on the materials, they were given an opportunity to discuss the texts. In one school, some pupils wrote their own versions of the text.

Of the four texts, Version B required the least teacher explanation during the work period, whilst pupils using Version A needed continual help with the content of the text. The table of results shown in Table 9.1 was produced in one of the schools.

Table 9.1

Version	No. who started	No. who finished	No. of times clarification sought
A	54	36	42
B	48	42	0
C	51	45	12
D	40	32	12

Some differences in the ways in which the pupils reacted to the different versions of the text became apparent during the work. These differences are now discussed.

Version A

Pupils were often uncertain of what they were expected to write down, and many asked 'Do I have to *do* anything here?' The questions 'How can they be represented? How would you show $\frac{1}{4}$, $\frac{1}{2}$, $\frac{3}{4}$ and other fractions?' did not generally elicit a written response, although some pupils asked what they had to do, and three pupils gave illustrations of the type shown in Figure 9.8, p. 120. The suggestion 'Perhaps you

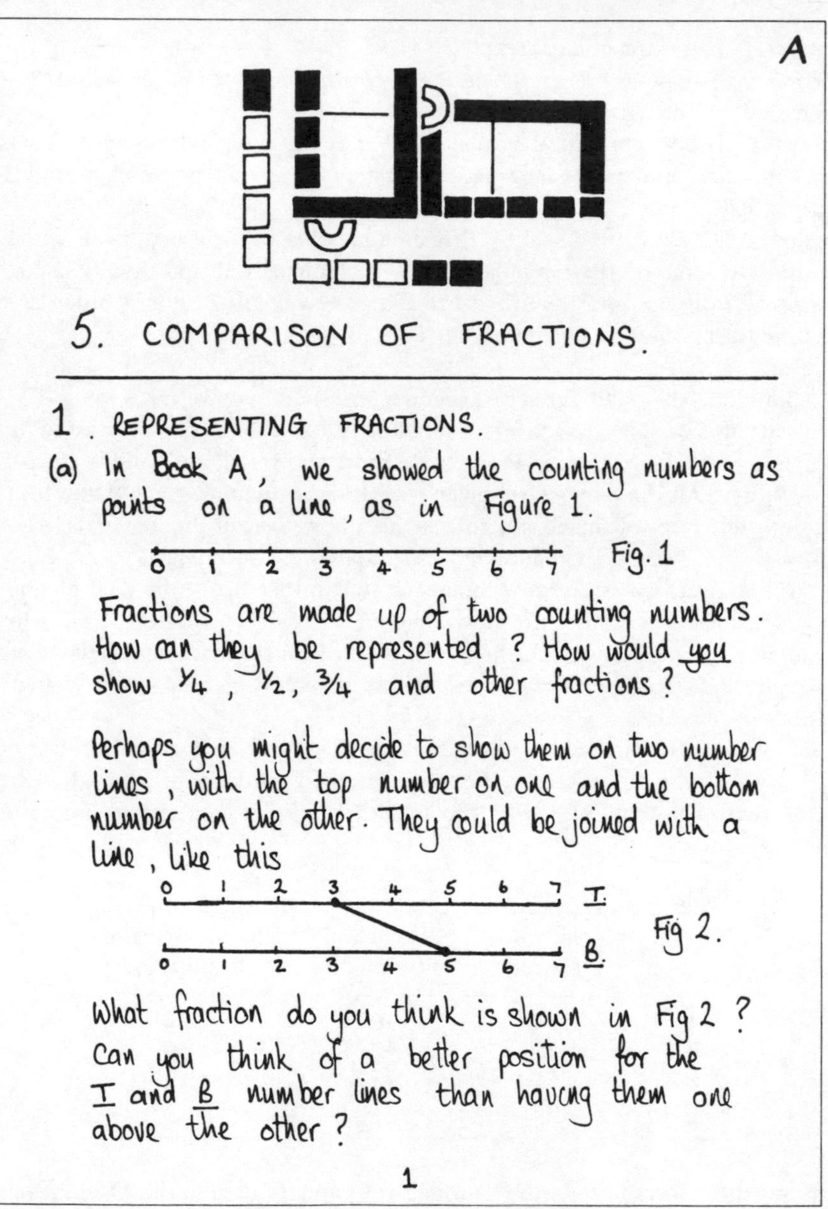

A

5. COMPARISON OF FRACTIONS.

1. REPRESENTING FRACTIONS.

(a) In Book A, we showed the counting numbers as points on a line, as in Figure 1.

0 1 2 3 4 5 6 7 Fig. 1

Fractions are made up of two counting numbers. How can they be represented? How would <u>you</u> show 1/4, 1/2, 3/4 and other fractions?

Perhaps you might decide to show them on two number lines, with the top number on one and the bottom number on the other. They could be joined with a line, like this

0 1 2 3 4 5 6 7 T.

Fig 2.

0 1 2 3 4 5 6 7 B.

What fraction do you think is shown in Fig 2? Can you think of a better position for the T and B number lines than having them one above the other?

1

Figure 9.4 (part) Version A, page 1 (reduced to 75%).

might decide . . .' led many children to copy Figure 2 of the text, although they were not asked to do so. Very few were sure whether they were required to answer the question 'Can you think of a better position for the *T* and *B* number lines?' The few who did answer simply said 'No'! The other pupils may have been baffled, or perhaps they considered that the question did not call for an answer. One pupil understandably asked, 'Do I have to do anything on this page?'

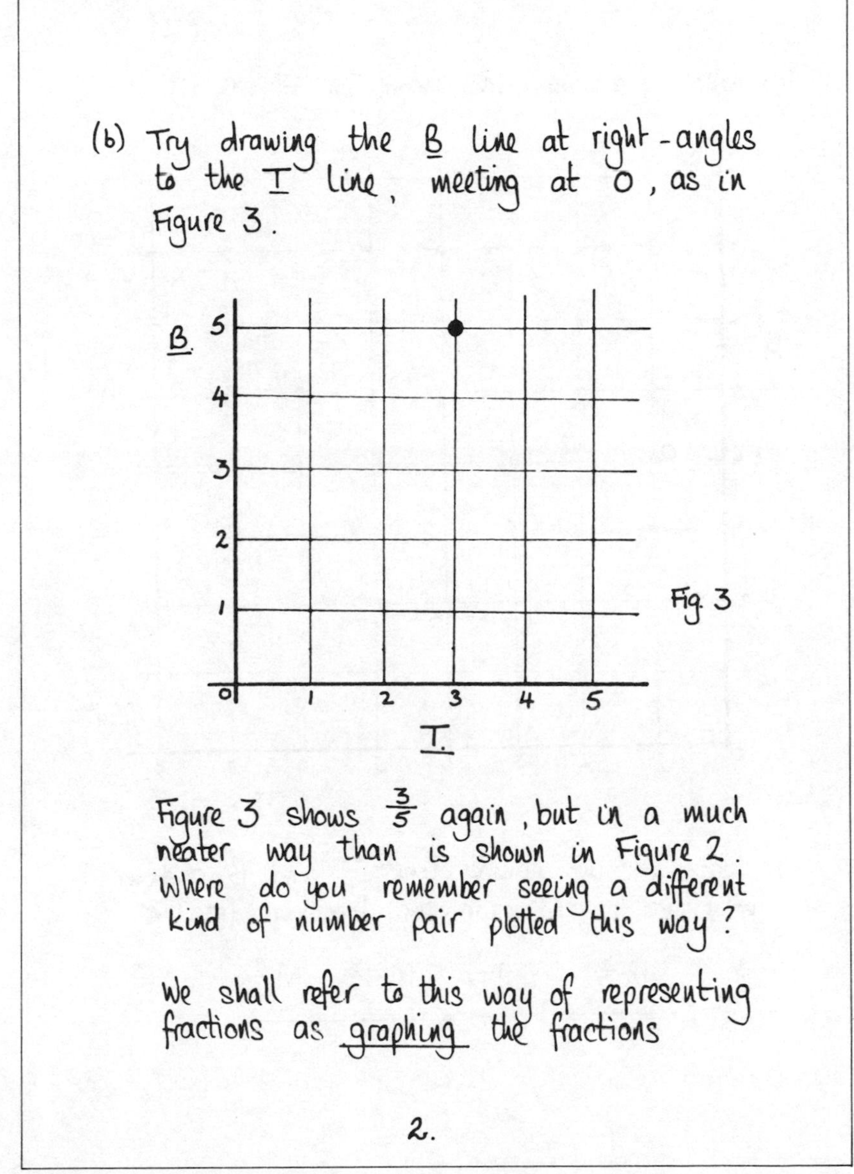

Figure 9.4 (contd.) Version A, page 2 (reduced to 75%).

Most pupils copied Figure 3, although it would seem that this was not intended. Several got the axes mixed up, and used the horizontal axis for *B*. Only a very few were able to answer 'Where do you remember seeing a different kind of number pair plotted in this way?'; again it is unclear what kind of response is required. The purpose of much of the information on page 1 was not clear to many pupils, and many asked what it had to do with the exercise. Finally, one pupil commented on this version that it was 'easy maths made difficult by funny writing'.

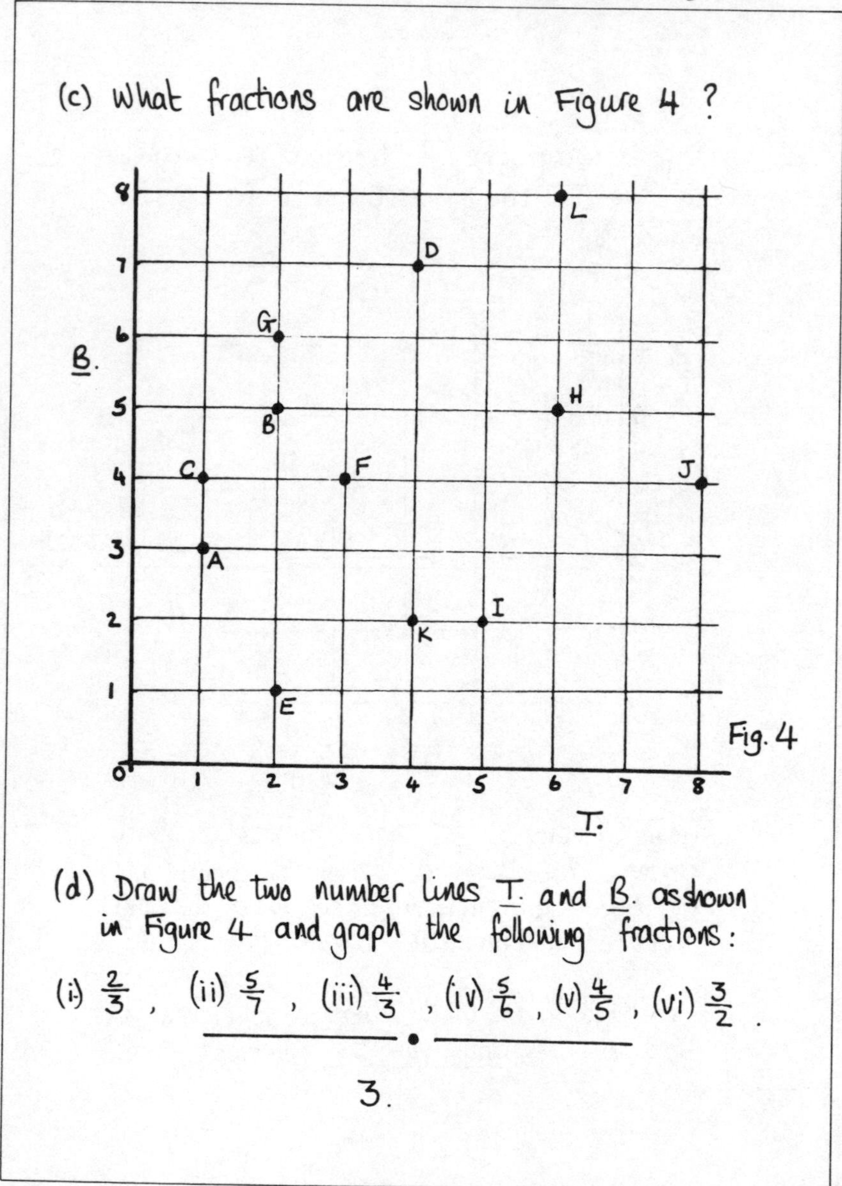

Figure 9.4 (contd.) Version A, page 3 (reduced to 75%).

Version B

This version caused much less questioning from the pupils than the other versions. Most pupils completed Question 1 successfully, and the answers were clearly set out. Children copied diagrams only when they were asked to do so. The order of

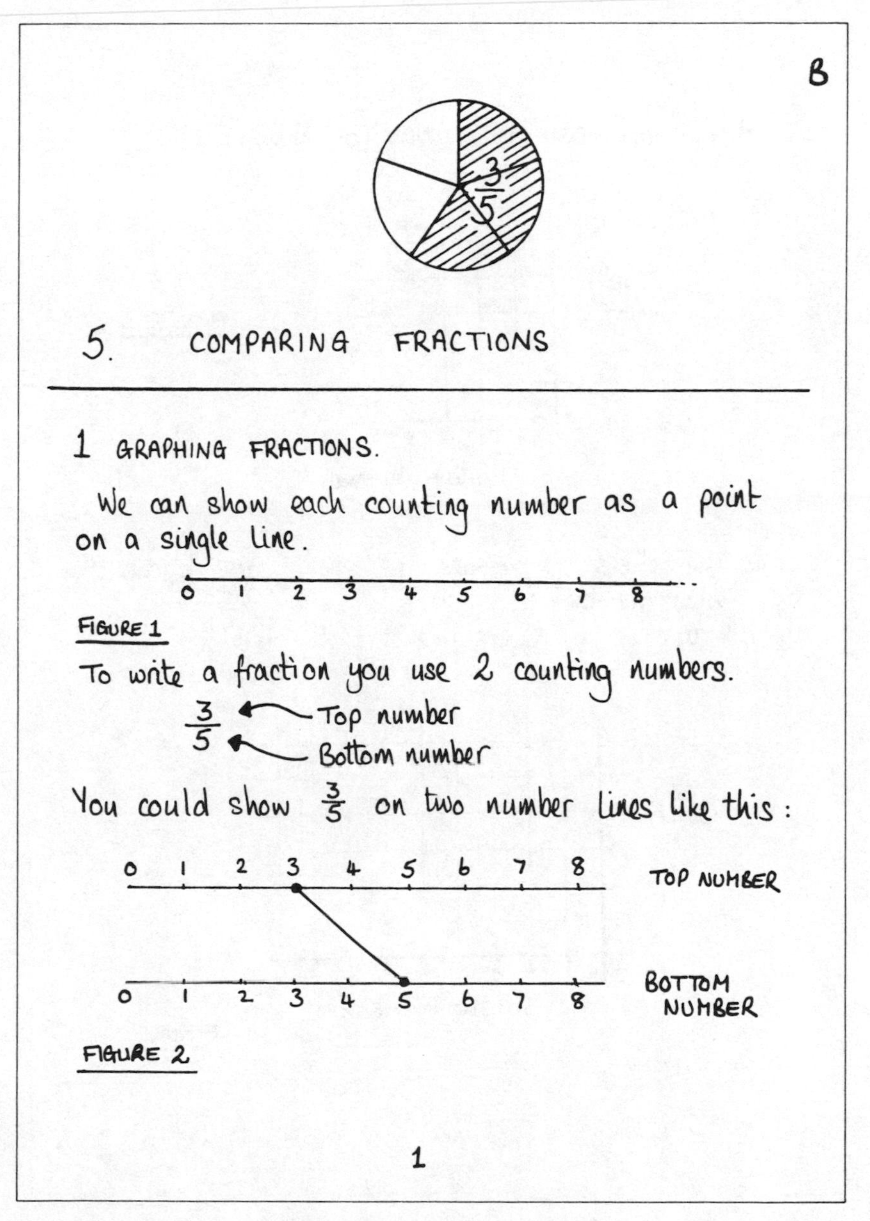

Figure 9.5 (part) Version B, page 1 (reduced to 75%).

the axes was different from that in Version A; the 'top number' was placed on the vertical axis, and 'bottom number' on the horizontal axis. This caused some confusion – one child, who remembered plotting coordinates, commented, 'I thought this was a bit confusing when they did the numbers the wrong way round

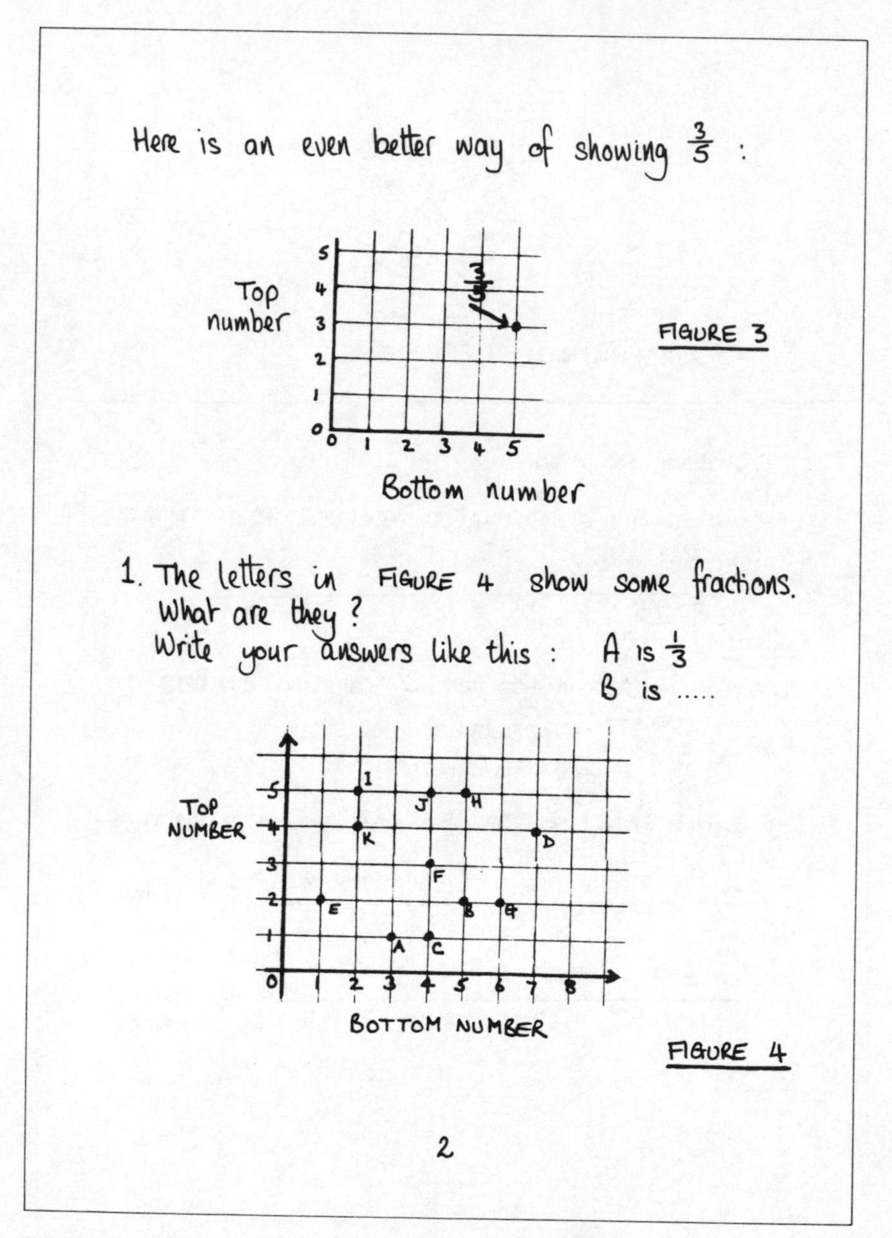

Figure 9.5 (contd.) Version B, page 2 (reduced to 75%).

than we were used to doing.' Plotting the point corresponding to $\frac{5}{6}$ caused a query from one child because the axes shown in the grid only went up to 5. Answers to Question 2 often consisted of letters scattered over the diagram, without the corresponding points being marked.

2. Make a copy of FIGURE 3.

Put letters on your grid for these fractions

P is $\frac{2}{3}$, Q is $\frac{5}{6}$, R is $\frac{4}{3}$, S is $\frac{4}{5}$

T is $\frac{3}{2}$, U is $\frac{3}{6}$, V is $\frac{1}{6}$.

Figure 9.5 (contd.) Version B, page 3 (reduced to 75%).

Version C

The pupils' questions gave their teachers the impression that many pupils only scanned page 1 as they looked for the first thing they had to do; they started the exercise on page 3 without fully reading the first two pages. The 'Read this' and 'Do this' statements did not split the work into clearly defined tasks in the way that was intended. Some pupils copied every diagram, whether or not this was required.

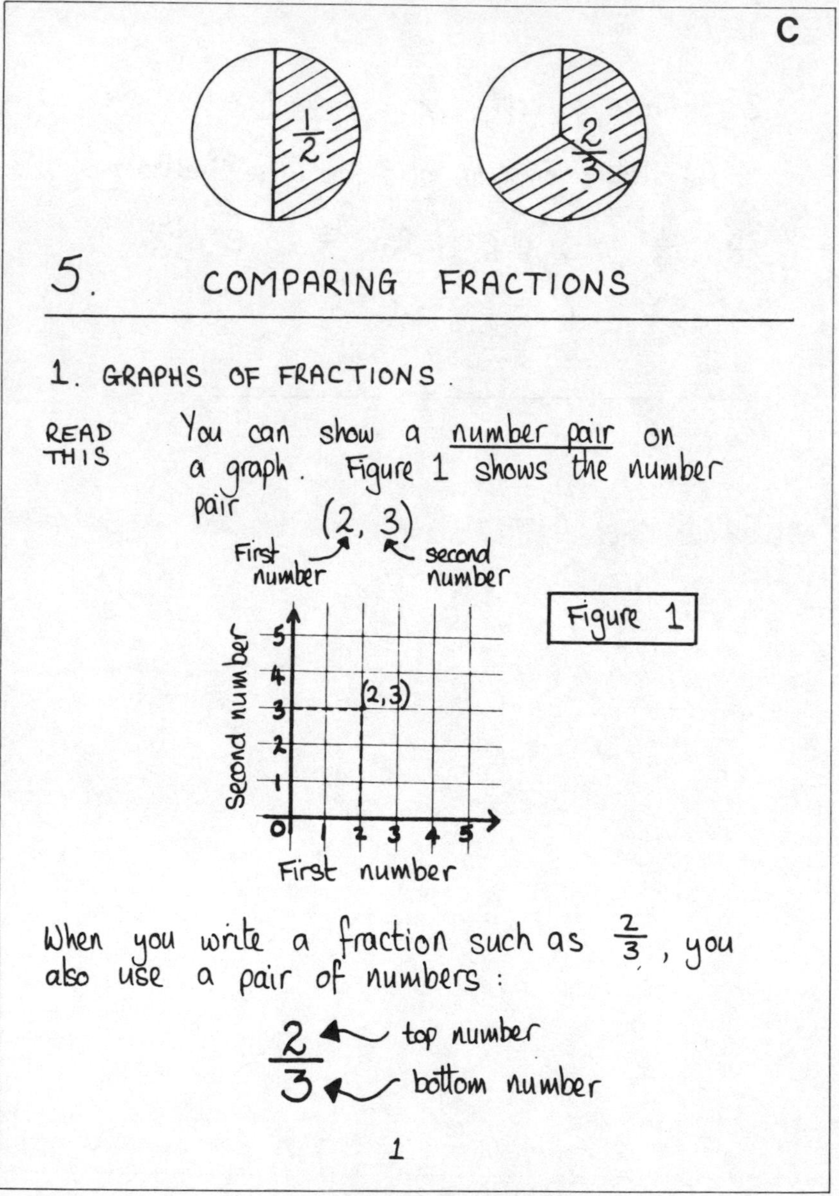

Figure 9.6 (part) Version C, page 1 (reduced to 75%).

Question 1 was not as well done in this version as in Version B. The instruction:

Write your answers like this: A is $\frac{1}{3}$ (Version B)

seems to be clearer than:

For example, A = $\frac{1}{3}$ (Version C)

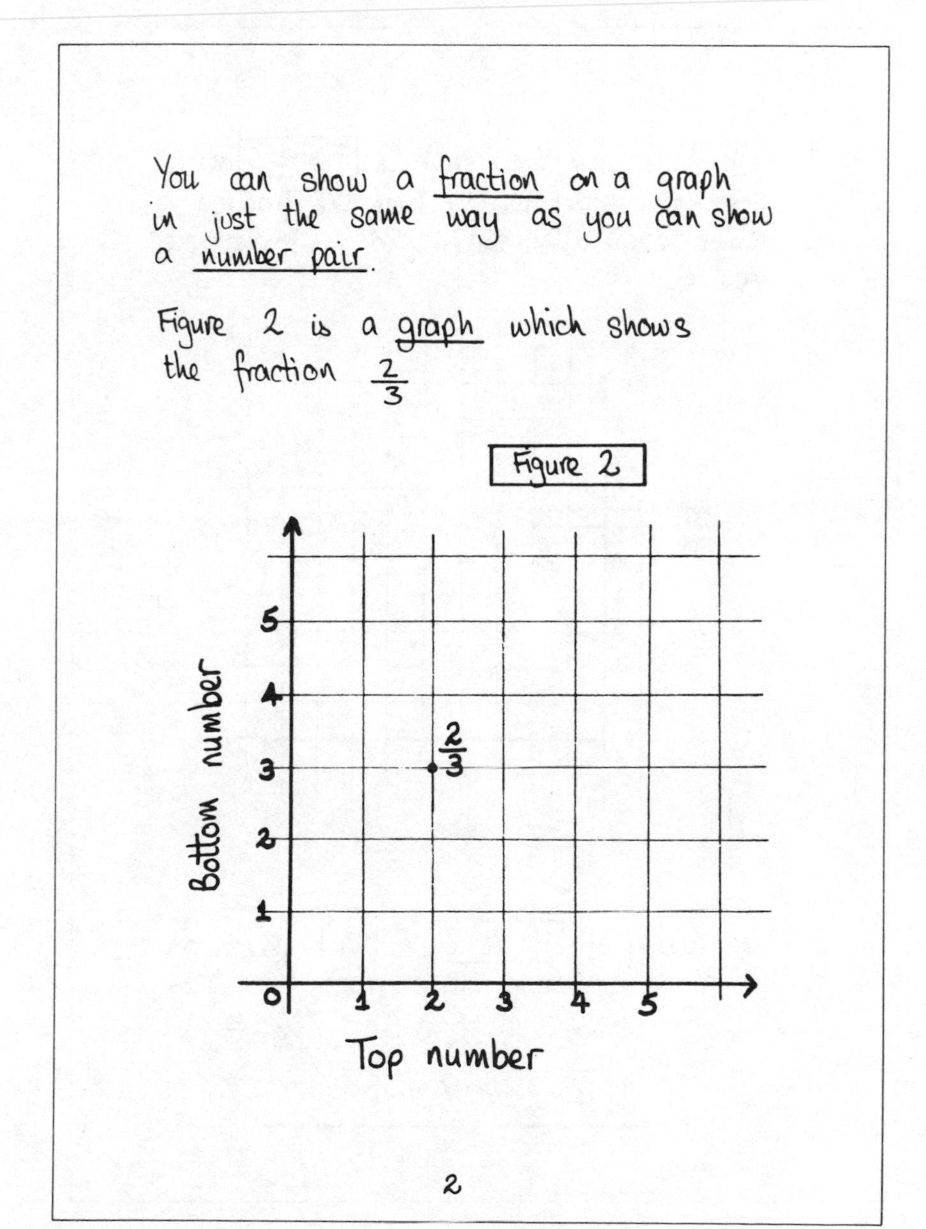

Figure 9.6 (contd.) Version C, page 2 (reduced to 75%).

Question 2 also was not done as well as in Version B: the dotted lines of Figure 1 often appeared on the pupils' grids, in spite of the fact that Figure 3 gave an example of what was anticipated. One child complained that 'the lettering was not clear but apart from that it was easy'. Another said, 'I thought the sheet was difficult to read, but thought the work was quite easy.'

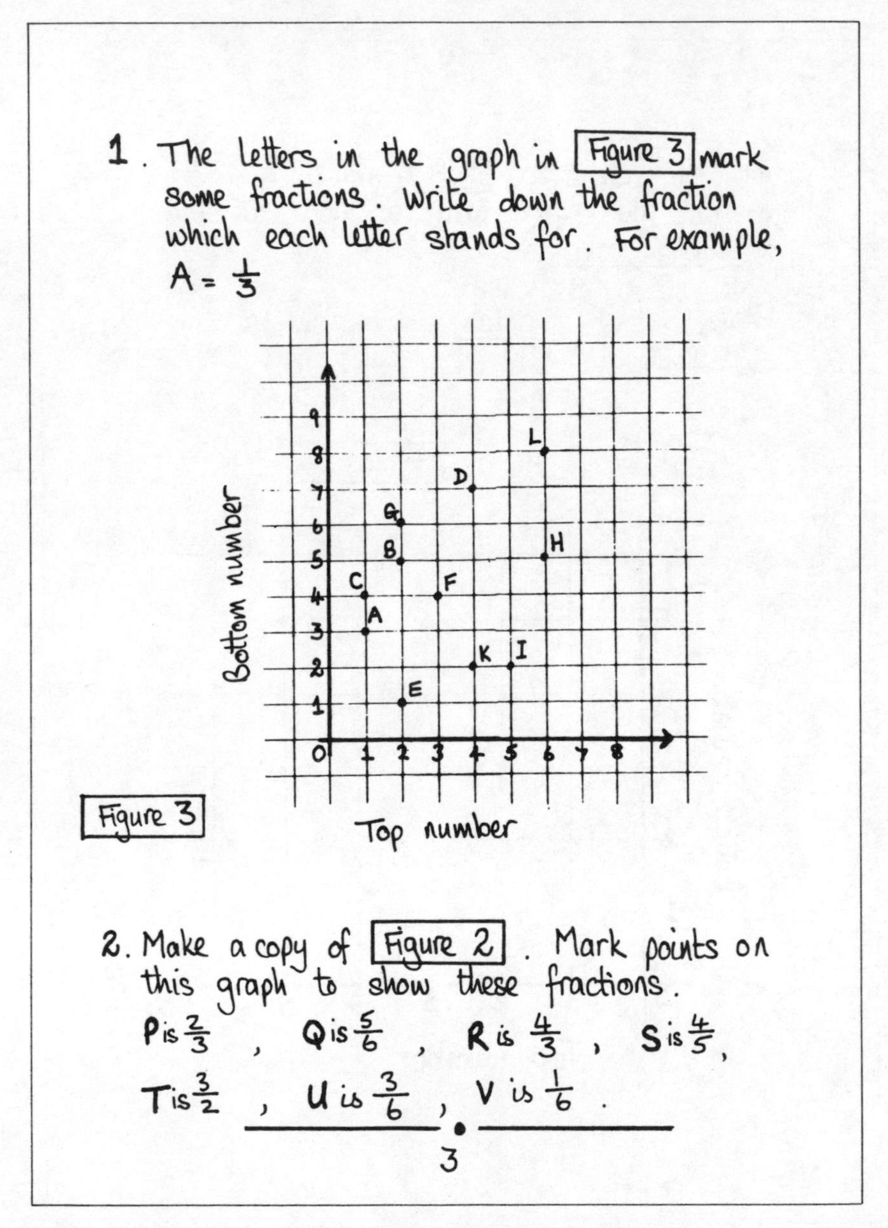

Figure 9.6 (contd.) Version C, page 3 (reduced to 75%).

Version D

This more verbose version was accepted without difficulty by the pupils in one of the schools; they read the first two pages and readily accepted the format. In the other school, however, this version produced the most outspoken criticism –

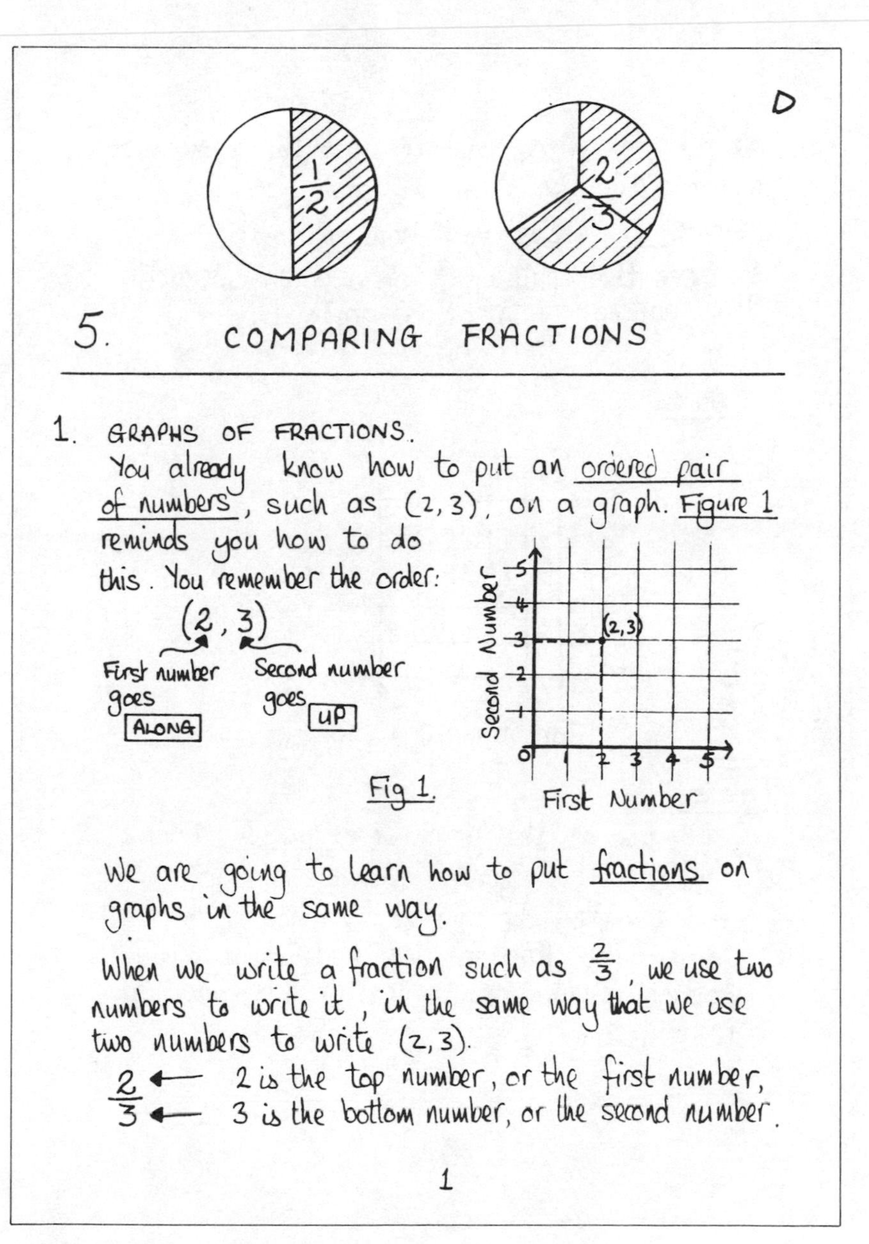

Figure 9.7 (part) Version D, page 1 (reduced to 75%).

perhaps because several pupils worked through both Version C and Version D. One of these children commented, 'I thought they were both easy. I don't think the title suits the work.' Here the fault is ours; the part of the chapter which the pupils worked on did not extend as far as the comparison of fractions. But the comment shows that some pupils *do* read the title of the work! The attitude of

We put $\frac{2}{3}$ on a graph in just the same way as we put (2,3) on a graph.
Figure 2 shows you how to do it.
You see that putting $\frac{2}{3}$ on a graph is exactly like putting (2,3) on a graph.

Fig 2.

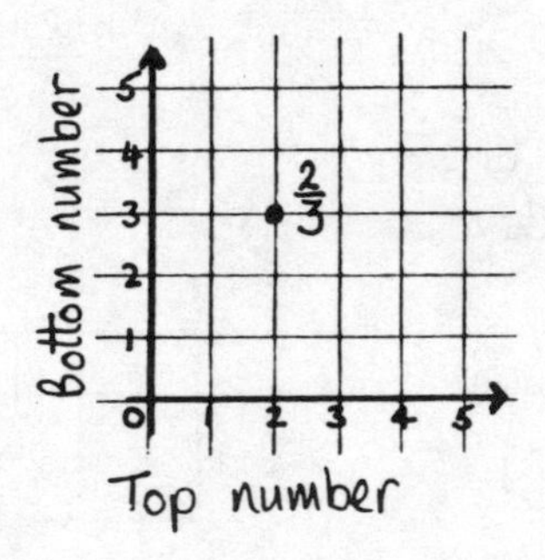

DO THESE

1. Look at the graph in Figure 3. Each letter in this graph shows a fraction. For example, the letter A shows where we put the fraction ⅓. Write down the fraction which each letter stands for, like this: $A = \frac{1}{3}$

2

Figure 9.7 (contd.) Version D, page 2 (reduced to 75%).

the pupils who did both versions was summarised by one of them as 'Too much words make it a bit complicated'; again the feeling came over that the work was easy but the reading difficult. Another pupil said, 'I have done C and D and they were the same but D had more writing in it and got me confused, while C has less writing and I thought it was better.'

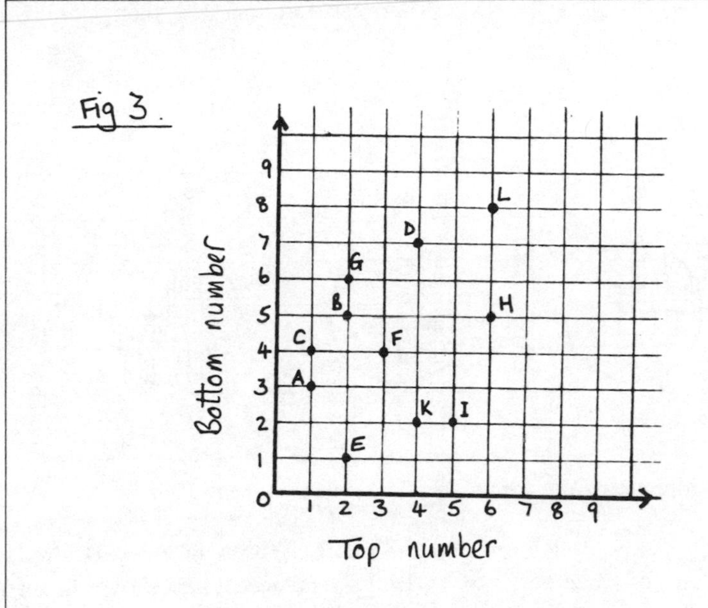

2. Draw axes and label them to show the top number and the bottom number, just as we did in Figure 2. Mark in a point on this graph to show each of these fractions. Your graph will look rather like Figure 3.

P is $\frac{2}{3}$, Q is $\frac{5}{6}$, R is $\frac{4}{3}$, S is $\frac{4}{5}$,

T is $\frac{3}{2}$, U is $\frac{3}{6}$, V is $\frac{1}{6}$.

3

Figure 9.7 (contd.) Version D, page 3 (reduced to 75%).

Further comments on the passage of secondary text

When the pupils had worked the text, members of the writing group discussed it with the teachers involved, and these discussions brought up further points for consideration. It was pointed out that there was only one example before the exercise, and this introduced only a fraction less than 1, whereas the exercise also asked the pupils to plot fractions greater than 1. The line $y = x$ could have been brought out more clearly as the divider between fractions less than 1 and those greater than 1.

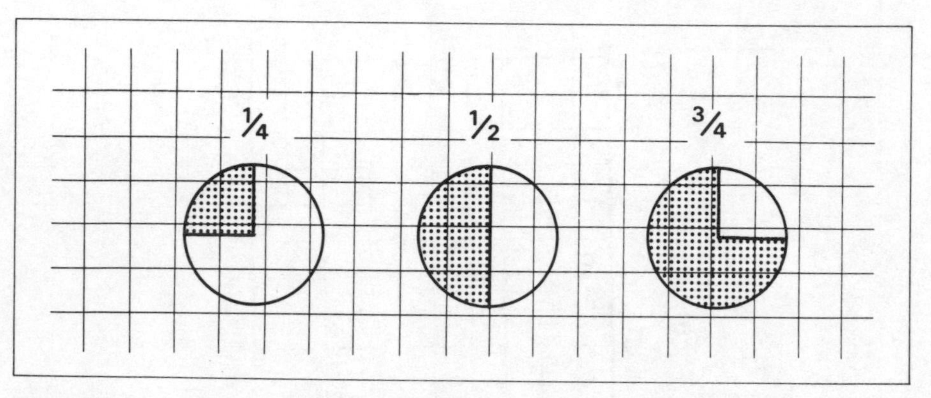

Figure 9.8 A response to Version A questions, see text, page 108.

In one school, some pupils produced their own versions of the text after the discussion, and one of these versions, by Paul Tyers, aged 11, is shown in Figure 9.9. In spite of his rather primitive spelling, Paul sees straight to the heart of the matter, omitting unnecessary introductory verbiage and realising that the text needs to show how to plot 'difficult' fractions such as $\frac{3}{1}$ and $\frac{1}{2}$; he also provides an 'activity' at the end of the exercise.

The following further points became clear to the writing group as the work progressed.

- The source text (Version A) reads as if a teacher had constructed it from the memory of a lesson which used teacher–pupil dialogue. The 'rhetorical questions' such as 'Can you think of a better way . . .?' are very suitable for class questioning, but they do not translate well into reading materials.
- The pupils thought the mathematical content was comparatively simple, but it was obscured by the difficulty of reading the text.
- In Versions A and B, Figure 2 acted as a 'red herring'. In a class lesson, this diagram might be used successfully by a teacher to exemplify one of several

Figure 9.9 (opposite) Paul's version of the text.

Paul Tyers

Fractions

A way of showing fractions is by useing cordints
Here is an extsample

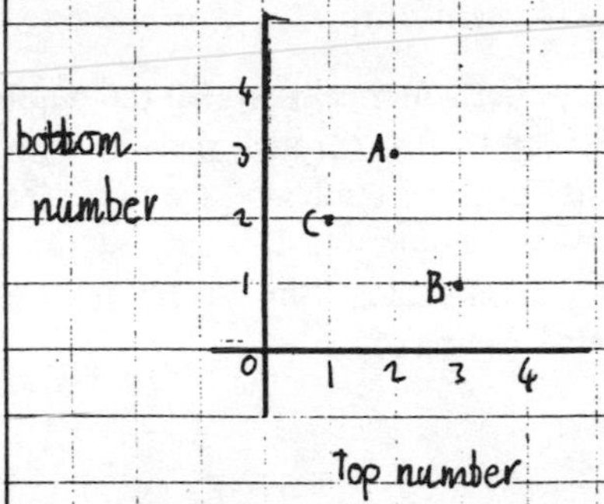

IMPORTANT
Rember to ladbel your axes correctly.

The fraction for piont A is $\frac{2}{3}$

The fraction for piont B is $\frac{3}{1}$

The fraction for piont C is $\frac{1}{2}$

1. Now find out these fractions
And write the answers in your book.

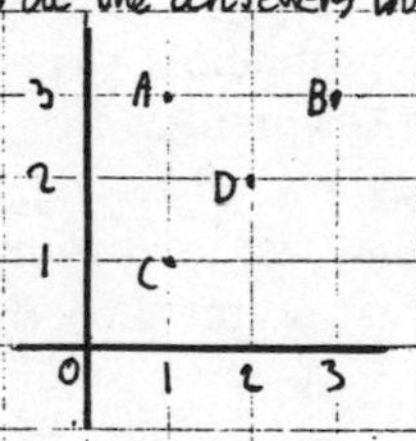

2 In your book draw a graph
and plot the pionts for these fractions
a $(\frac{1}{3})$ b ( ) c $(\frac{1}{2})$ d $(\frac{2}{8})$ e $(\frac{5}{10})$ f $(\frac{1}{10})$

3. draw a graph wih with an axes that goes up to 7 and plot thesse fractions to make a shape A $\frac{4}{5}$ B $\frac{6}{3}$ C $\frac{7}{5}$ D $\frac{7}{2}$ E $\frac{1}{2}$ F $\frac{1}{3}$ G $\frac{2}{3}$

approaches to illustrating fractions, but in the text it has to be presented to the pupil without this explanation. The text then immediately rejects this diagram, causing confusion to many pupils.

- The text needs to distinguish clearly between those parts which are purely exposition and those which require some written response. Some pupils copied Figure 2 of Version A, evidently feeling that some written activity on their part was called for.
- Pupils need to be taught the skills of reading mathematical text. All the pupils who worked on these versions had a tendency to rush through the reading to find something they had to *do*. To counteract this tendency, many writers keep verbal material in mathematical text to a minimum. It would be more fruitful if pupils could learn how to extract the meaning from written text, so that they could use it as a resource for their learning.

The primary passage

The work with secondary pupils preceded that with primary pupils, and the group therefore started work on a passage from a text for primary pupils with clearer ideas of the pitfalls and difficulties.

A short passage from *Mathematics for Schools* by Fletcher *et al.* was selected; teachers had told the group that pupils often experienced difficulty with the passage which introduces pupils to the use of the protractor. This passage, in Level II, Book 4, pp. 23–24 of the first edition (1971), was therefore chosen. Further reasons for the choice of a passage from 'Fletcher' were that this series of texts is widely used, and that revisions in the second edition (1980) were intended to make the task of reading easier for the pupils.

The passages in both editions which introduce protractors were analysed for flow of meaning, using the techniques described in Chapter 6. In the light of this analysis, two members of the group produced a single revised version. This version, together with the original versions from both editions, was used with children. All three versions were produced in the same typed format, so that they should look as similar as possible. They are:

Version A. Level II, Book 4, first edition, pp. 23–24.
See Figure 9.10, pages 124 and 125.

Version B. Level II, Book 4, second edition, pp. 22–23.
See Figure 9.11, pages 126 and 127.

Version C. The rewritten version.
See Figure 9.12, pages 128 and 129.

Two hundred and four children from six different primary schools worked the different versions of the texts. They were all aged 8–10 years. Some came from junior schools and some from middle schools, and all were pupils of teachers who were following a course for a Diploma in Mathematical Education. These teachers had done some work on reading and mathematics, and were all very interested in taking part in the work. They were given the following instructions.

1. Use classes of second- and third-year junior age (aged 8–10), who have done *some* work on angles. If necessary, a brief introduction to angular measure might be given before the children work the pages.
2. Please ensure that circular protractors marked at 10° intervals are available for each child.
3. Distribute the three versions randomly, so that each child has one version.
4. Ask the children to take *any* problems or queries to *you* and not to each other.
5. Place a mark on the script every time each child asks for help or clarification.
6. Allow the children about ¾ hour to do the work.
7. Encourage the children to comment on the scripts after the session. Any general reactions and comments from the children – and from you – would be a great help!

The results obtained are shown in Table 9.2.

Table 9.2

Version	No. who started	No. who finished	No. of times clarification sought
A	70	38	193
B	64	34	221
C	70	59	121

Comments on the three versions of the passage of primary text

Version A

The children's reactions to this passage were of the type 'It looked boring' and 'It was ever so hard'. Although 38 children 'finished', none of them attempted questions 2 and 4 on page 2. On page 1, the children found difficulty with the notation BAF and MAF for angles. Nearly all of them asked about this, and some wrote 'right angle' and 'straight line' instead of 90° and 180° to complete the sentences in question 1. Question 2 also produced difficulties, as some children found the arms of the angles too short for their protractors. One child was found trying to measure the ∠ sign in ∠ BAD. The text was in worksheet form rather than in the form of a textbook page, and the boxes provided for the answers were also too small – our mistake.

It seemed that in answering question 1 on page 2, many children read the first instruction 'Use a protractor . . . ', measured the angles and filled in the table, and then finally wrote a number near the measured size of the angle into the 'estimate'

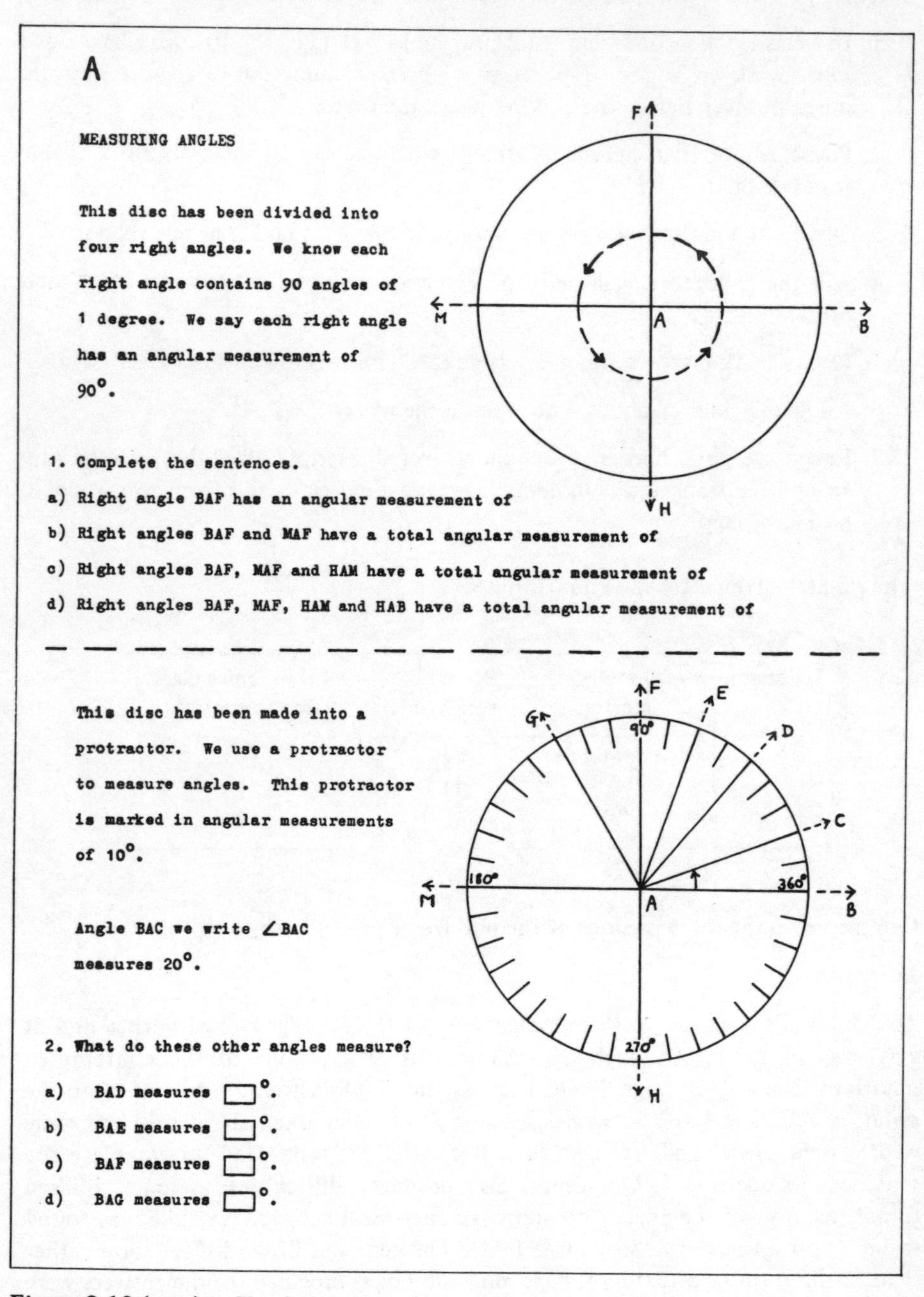

A

MEASURING ANGLES

This disc has been divided into four right angles. We know each right angle contains 90 angles of 1 degree. We say each right angle has an angular measurement of 90°.

1. Complete the sentences.

a) Right angle BAF has an angular measurement of

b) Right angles BAF and MAF have a total angular measurement of

c) Right angles BAF, MAF and HAM have a total angular measurement of

d) Right angles BAF, MAF, HAM and HAB have a total angular measurement of

This disc has been made into a protractor. We use a protractor to measure angles. This protractor is marked in angular measurements of 10°.

Angle BAC we write ∠BAC measures 20°.

2. What do these other angles measure?

a) BAD measures ☐°.

b) BAE measures ☐°.

c) BAF measures ☐°.

d) BAG measures ☐°.

Figure 9.10 (part) Version A, page 1.

line of the table. Those children who reached question 3 answered it quite well, but were uncertain whether to write down the *size* of an angle (e.g. 40°) or its *name* (e.g. a)) in the list of acute and obtuse angles.

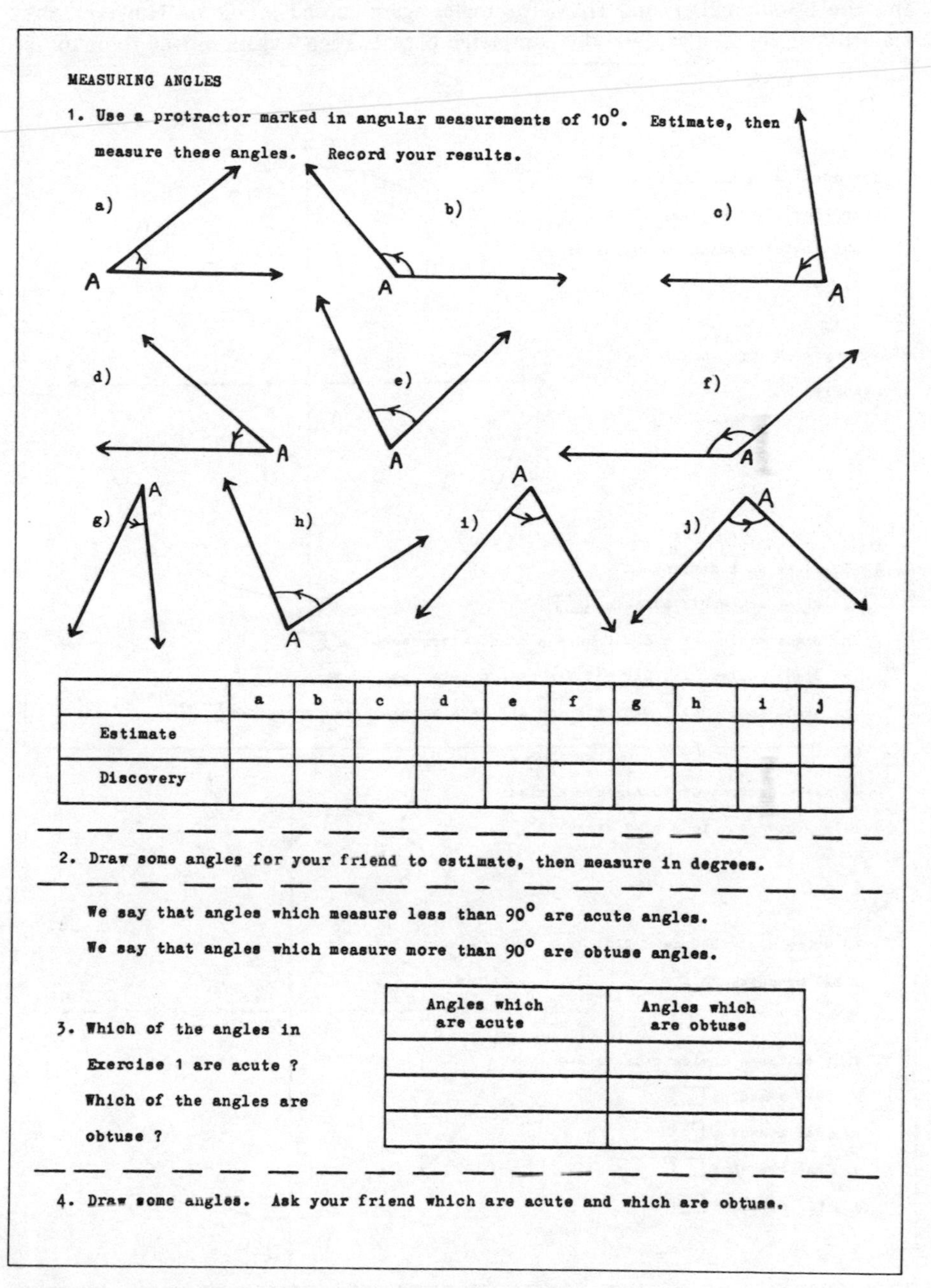

MEASURING ANGLES

1. Use a protractor marked in angular measurements of 10°. Estimate, then measure these angles. Record your results.

a) A b) A c) A d) A e) A f) A g) A h) A i) A j) A

	a	b	c	d	e	f	g	h	i	j
Estimate										
Discovery										

2. Draw some angles for your friend to estimate, then measure in degrees.

We say that angles which measure less than 90° are acute angles.

We say that angles which measure more than 90° are obtuse angles.

3. Which of the angles in Exercise 1 are acute ? Which of the angles are obtuse ?

Angles which are acute	Angles which are obtuse

4. Draw some angles. Ask your friend which are acute and which are obtuse.

Figure 9.10 (contd.) Version A, page 2.

Version B

In this version, taken from the second edition, the text contains fewer words, but more than half the children asked the meaning of 'partitioned' and 'complete', and the notation BAF and HAM for angles again caused problems. However, the majority of children successfully completed page 1. Page 2 again proved to be more

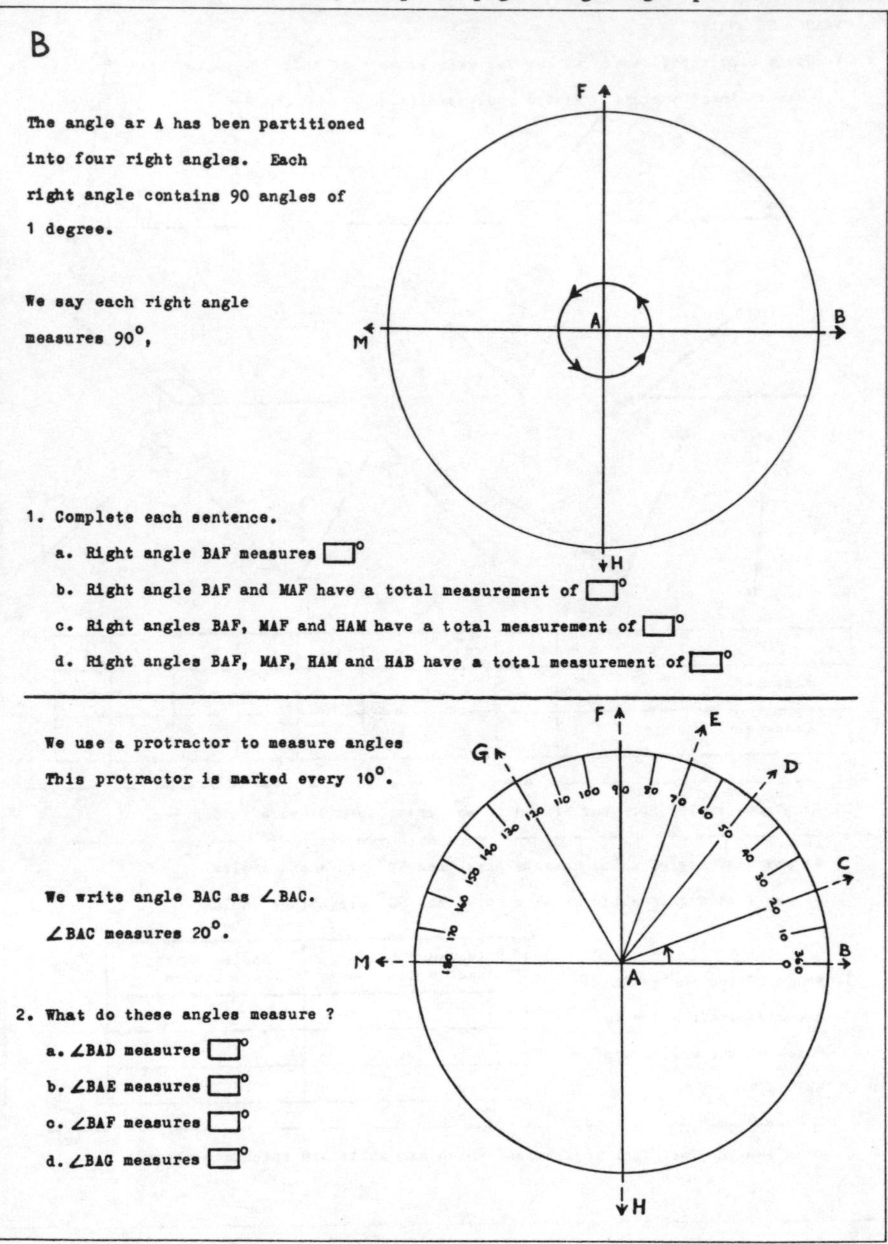

B

The angle ar A has been partitioned into four right angles. Each right angle contains 90 angles of 1 degree.

We say each right angle measures 90°,

1. Complete each sentence.
   a. Right angle BAF measures ☐°
   b. Right angle BAF and MAF have a total measurement of ☐°
   c. Right angles BAF, MAF and HAM have a total measurement of ☐°
   d. Right angles BAF, MAF, HAM and HAB have a total measurement of ☐°

We use a protractor to measure angles
This protractor is marked every 10°.

We write angle BAC as ∠BAC.
∠BAC measures 20°.

2. What do these angles measure ?
   a. ∠BAD measures ☐°
   b. ∠BAE measures ☐°
   c. ∠BAF measures ☐°
   d. ∠BAG measures ☐°

Figure 9.11 (part) Version B, page 1.

difficult, but the changed text of question 1, 'Estimate and then measure', ensured that the table was completed in the intended order – when 'estimate' had been explained. The labelling of each angle in question 1 by a letter within the angle also ensured that the names of the angles were recorded in the table of question 2. On the whole, too, version B was completed more quickly than Version A.

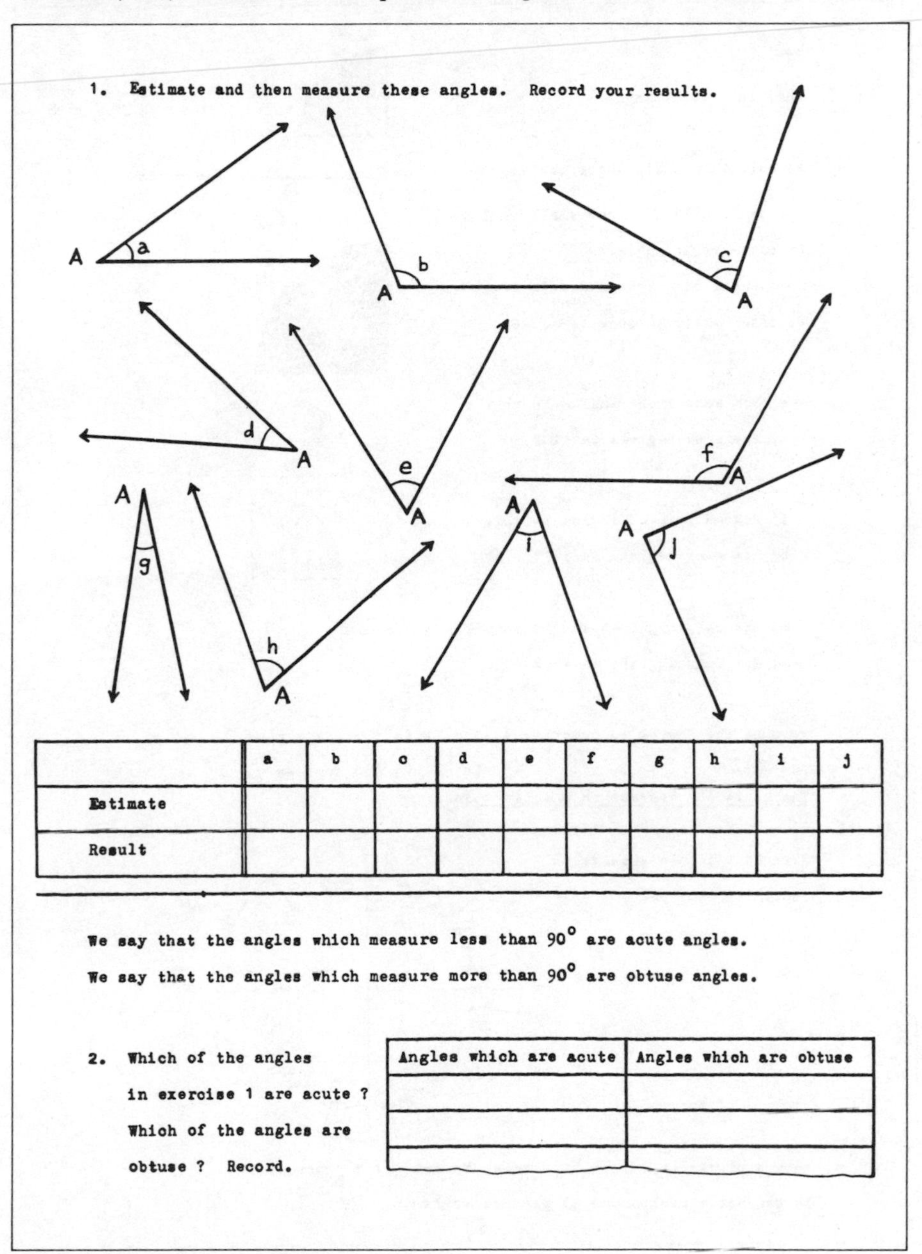

1. Estimate and then measure these angles. Record your results.

	a	b	c	d	e	f	g	h	i	j
Estimate										
Result										

We say that the angles which measure less than 90° are acute angles.

We say that the angles which measure more than 90° are obtuse angles.

2. Which of the angles in exercise 1 are acute ? Which of the angles are obtuse ? Record.

Angles which are acute	Angles which are obtuse

Figure 9.11 (contd.) Version B, page 2.

Version C

More children finished this version than the others, there were only about half as many queries, and on the whole the children's reactions were more favourable. Although they knew what they wanted to write in response to the questions on

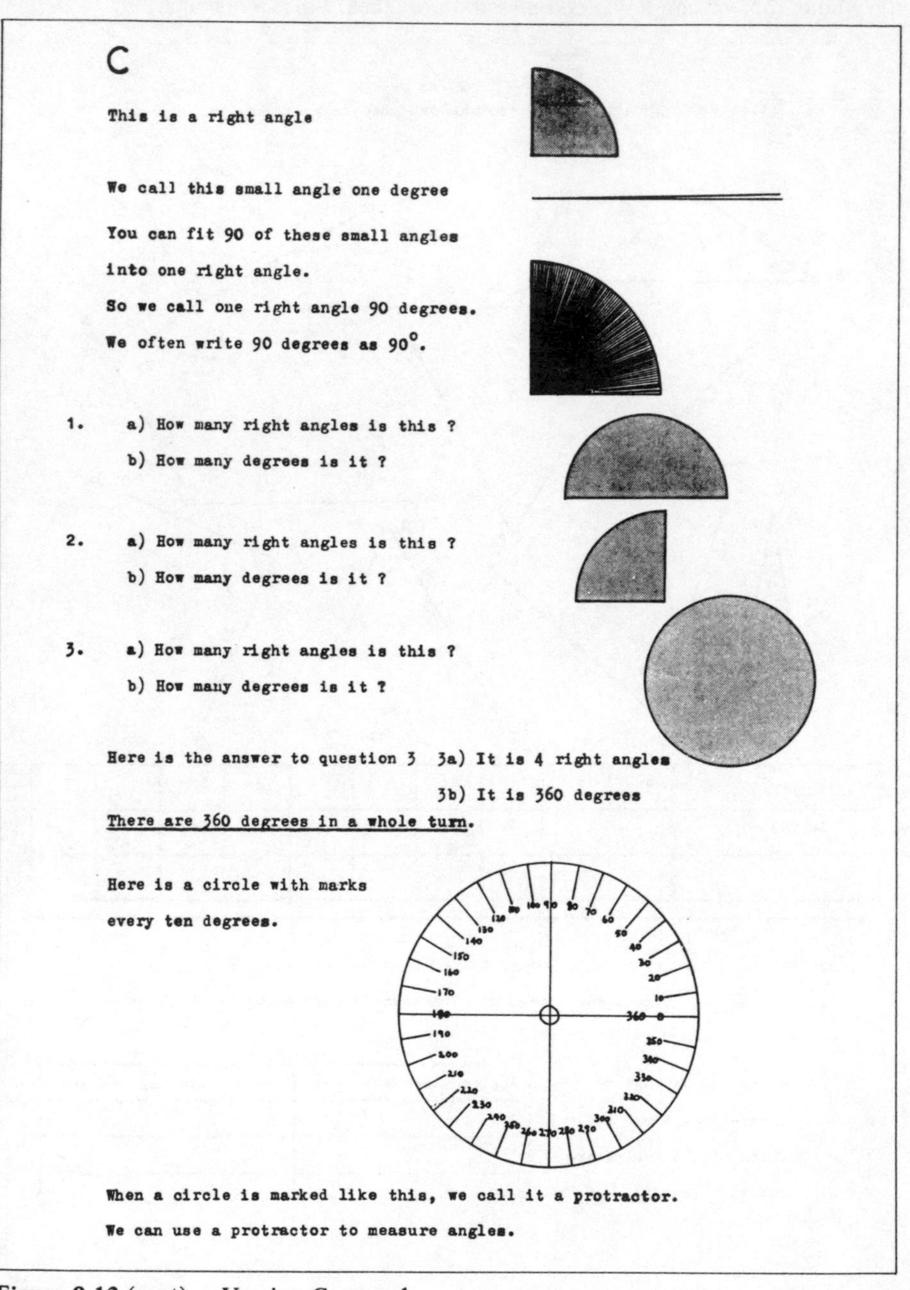

C

This is a right angle

We call this small angle one degree

You can fit 90 of these small angles
into one right angle.
So we call one right angle 90 degrees.
We often write 90 degrees as 90°.

1. a) How many right angles is this ?
   b) How many degrees is it ?

2. a) How many right angles is this ?
   b) How many degrees is it ?

3. a) How many right angles is this ?
   b) How many degrees is it ?

Here is the answer to question 3 3a) It is 4 right angles
3b) It is 360 degrees

<u>There are 360 degrees in a whole turn</u>.

Here is a circle with marks
every ten degrees.

When a circle is marked like this, we call it a protractor.
We can use a protractor to measure angles.

Figure 9.12 (part) Version C, page 1.

page 1, many children asked where they should write the answers.

Page 2 produced more comments than queries, and the teachers thought that most children could follow the text but were distracted by the novel approach. Some children measured the angles in the 'comic strip', and it also became clear

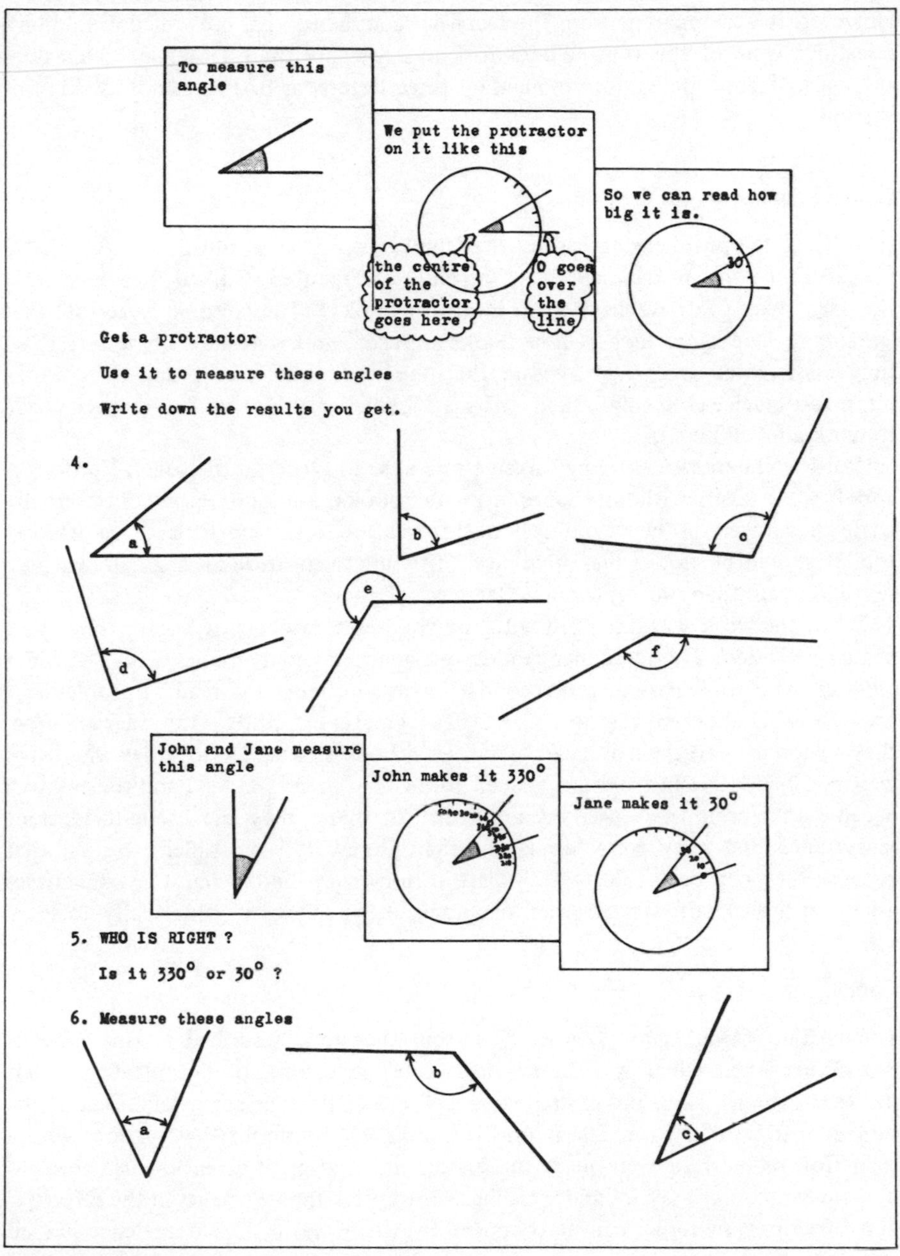

Figure 9.12 (contd.) Version C, page 2.

that the arms of the angles in question 4 were too short for the protractors many of the children were using. In spite of this, however, many children measured the angles with satisfactory accuracy.

The texts issued to the children were in worksheet form, but version C contained no spaces for the children to write their answers, and this produced many queries. However, it was apparent from the teachers' comments that most of the children's questions were of the type 'Where do I do it?' rather than 'How?' or 'What does this mean?' The labelling of an angle by three letters, as BAF, was not used in this version.

General comments on the primary passage

In spite of the problems caused by the 'comic strip' style of illustration, Version C proved to be the most acceptable. One child commented, 'I liked that once I got the idea. Why can't we have them in proper books?' The children were told that they were helping to look at how books are used, and were not 'doing a test', so they commented freely. Many said that the work was boring, and that 'books jump too much'. One child asked, 'Why are books easy and boring, then you have to work a lot and it's hard?'

Many children said that they liked to be able to write on the pages. Versions A and B were taken without change from textbooks, and children would normally write the answers in their exercise books. In relation to this particular piece of work, too, it should be noted that it is more difficult to measure an angle drawn on a textbook page than one drawn on a flat sheet of paper.

When the work was discussed with the children's teachers, it became clear that primary children shared similar reading problems in mathematics with the older children who took part in our secondary work, but that the reading problems of these younger children were more acute. To a greater extent than the older children, they read one instruction and carried it out without looking ahead to see whether it was modified later. Instructions needed to be very precise indeed and the children became anxious if there was any ambiguity in them; they then sought constant reassurance that they were 'doing the right thing'. It is possible, however, that because they had been told to seek clarification from the teacher, they sometimes asked for the sake of asking, rather than going ahead as they would usually do.

Conclusion

The writing group learnt a great deal from the work described in this chapter. Examining a text which is to be rewritten encourages one to concentrate on what the text actually says, rather than taking the usual rather superficial look at the subject matter of a page. The pupils' reactions left no doubt that versions with a clear flow of meaning were more successful, and that great attention also needs to be paid to details of layout and wording. Substantial improvements in the effectiveness of written materials can be produced by these means. Our experience has led us to think that work of this type is a valuable exercise in in-service education for teachers, and that teachers' ability to write clear materials for their pupils can be improved by it.

Of course, no teacher ever relies entirely on written materials for the teaching of mathematics, and the successful integration of written materials into the variety of learning experiences arranged by the teacher is of great importance. In the next chapters, we discuss how the teacher may work towards producing effective written materials, and how the pupils may be encouraged to improve their reading skills in mathematics.

10 Improving written materials

Introduction

In this chapter we suggest ways in which written materials can be made clearer and easier for pupils to use. Every teacher of mathematics at some time writes his own materials, either on work-cards or on worksheets for duplication, and many teachers also write test and examination questions in mathematics. Hence these suggestions are addressed to general primary teachers as well as to specialist teachers of mathematics in secondary schools. We hope, too, that writers of materials for publication will also consider our work, because our study has shown that published work in mathematics for children is not always easy for them to read.

We give brief advice and guidelines for the teacher, based on the features of mathematical text discussed in previous chapters.

Vocabulary and syntax

Mathematics books often contain difficult vocabulary; in addition to the technical jargon of mathematics, difficult OE (Ordinary English) words are often used and are combined in complex sentences. Some recent books, however, claim to have 'simple vocabulary'. Such claims should be treated with caution, because examination of the text sometimes shows that the problem has been solved by avoiding the use of words as much as possible – the books are largely compilations of exercises written in symbols. In few textbook series is there an attempt to teach vocabulary systematically; usually only a few definitions are given, and few books contain glossaries. Especially at primary level, a glossary can clarify the meaning of the technical words used (and its compilation may also help the writer to control his vocabulary). In definitions and glossaries, it is important that the language used to explain the meanings of terms should be simple.

Many mathematical writers have acquired from their own education a formal style of writing; this is out of place in text for pupils. A well-intentioned attempt to replace formal language by clear explanations can unwittingly lead to verbosity and increased reading difficulty. For example, the second of a pair of parallel primary texts, *Alpha Mathematics* and *Beta Mathematics,* was written because many teachers found the first only suitable for above-average pupils. However, *Beta,* written to cater for rather less able pupils, and intended to be clearer, has a higher measured reading age than *Alpha* because it uses longer sentences and more polysyllabic words. Hence, writers of mathematics materials need to be aware of the importance, and the difficulty, of writing in simple language.

This difficulty is compounded by the fact that very little mathematical vocabulary is used in the children's everyday conversation. Earp and Tanner (1980) point to the necessity of 'making mathematical language a part of children's lives'. An example of a published series of texts which try explicitly to teach mathematical vocabulary is the series of *Bronto Books* for young children which accompany the *Nuffield Maths 5–11* scheme (1979). Many infant teachers and their pupils have found these simple reading books valuable and enjoyable. In later work in mathematics, too, vocabulary needs to be carefully introduced and reinforced.

The following guidelines may be found helpful in controlling the vocabulary and syntax of mathematical writing for children.

- When a new word is introduced it should be used several times, in different contexts if possible.

- New vocabulary should be introduced only if the word is really necessary; words which appear only on one occasion should be carefully reviewed. For instance, on a single page of a primary book, 'table', 'chart' and 'matrix' may all be used to describe the same thing. It would be better to decide which one of these words to use, and to use it consistently.

- There should be a controlled use of vocabulary: the introduction of new words should be planned, so that a limited number of new words are introduced at any stage. Ways should be found of reviewing the meanings of words from time to time in the text.

- Some verbs have as specialised a usage in ME as do nouns; for example, the use of 'solve' in 'solve these equations' is a specialised ME usage. These verbs also need careful introduction.

- A glossary is a helpful tool for readers. This may be restricted to words which are new in a particular book, or it may be cumulative to a series. When a teacher is writing his own materials, he can make a glossary at the same time for the pupils to refer to.

- The present tense should be used when possible, and the conditional mood avoided. Words such as 'if', 'suppose', 'given that' are traditional openings in mathematical problems. They are largely unnecessary. For example, the question:

    Given that butter costs 47p a block, what would be the cost of 5 such blocks?

  can be replaced by:

    Butter costs 47p a block. How much do 5 blocks cost?

- It is advisable to avoid sentences which require the reader to remember introductory clauses. For example:

    Draw a circle of radius 4.2 cm.

  is more readable than:

    Using a radius of 4.2 cm, draw a circle.

Similarly,

> Taking π as 3.14 and assuming that the tank is a perfect cylinder of height 95 cm and radius 25 cm, find the volume of water it will hold.

can be simplified to:

> A cylindrical tank is 95 cm high and has a radius of 25 cm. Find its volume. (Take $\pi = 3.14$.)

- Rhetorical questions should be avoided. These include 'Do you know another way of doing . . . ' and 'Can you think what is meant by . . . . ', where the type of response required is not clear to the pupil. Pupils may simply ignore these questions, or write down 'No'; neither of these responses is desirable.

Symbols

Symbols are always needed in mathematical text, but their excessive use can cause reading difficulties for pupils. Munro (1977) gave many examples of children who found difficulty in interpreting symbols, and called attention to a particular factor which could hinder pupils' interpretation of symbols – a lack of spatial awareness. The context also contributes much to the interpretation of symbols, as Skemp (1971) has emphasised.

Symbolism is used at different levels; initially a single new symbol is introduced and has a specific referent. Later, combinations of symbols are used, and have to be interpreted with specific reference to their spatial arrangement. For example, there is an important difference between $(-3)^2 - 1$ and $-(3^2) - 1$ and the pupil must interpret each set of symbols with regard to their spatial arrangement.

The following guidelines for the introduction of mathematical symbols in text may be found useful.

- When a new symbol is introduced, the pupil first needs to understand the concept symbolised and to become familiar with the way the symbol is used initially before that use is extended.
- Symbolic notation should not be used for its own sake; only symbols which are vital to the material should be used.
- All the symbols introduced in a text should be included in the glossary. Even the most commonly used symbol needs explanation, and not all pupils will pick up the idea it represents at its first introduction. Glossaries in texts for secondary pupils may need to include symbolism usually taught in primary schools.
- The introduction of symbols should be carefully planned; pupils should not be confronted by too many new symbols at one time.
- When symbols are combined, and their spatial arrangement contributes to the meaning, attention needs to be drawn to this.
- In clarifying the meaning of the symbols, care should be taken to avoid verbose text.

Graphic material

Graphic material is another important feature of mathematics text; it includes pictures, diagrams, graphs and tabulations. Pictures are more often found in text for young children, while diagrams, graphs and tables take over for older children.

Campbell (1978) points out that 'pictures are an important tool used in the teaching of primary mathematics'. She highlights the fact that children need to be taught to read these pictures if they are to perceive the mathematical relationships depicted in them. It is not always realised that older children also need to be taught to interpret the graphic material in their texts. This is a task for the teacher rather than one that can be met by the use of written materials, although the writer and the illustrator can help greatly by producing clear graphic material.

In Chapter 5, graphic material was classified as *essential, related* or *decorative*. Pupils may not find it easy to distinguish between essential and non-essential graphic material; and since they may learn from experience that it is not always necessary to read the diagrams, they may overlook essential graphic material. It is sometimes the case that published material includes diagrams whose decorative appeal obscures their essential mathematical purpose. Guidelines for the production of visual material are given below.

- Diagrams should be placed on the page so that they relate easily to the relevant part of the text.
- When a diagram is essential, the text should make this clear. The written material should help the reader to extract the essential points from the graphic material.
- Diagrams should be as clear and simple as possible, and the style of graphic material within a text should be consistent.

Text as a whole

In considering the text as a whole, attention needs to be paid to the *clarity* and *flow of meaning* of the whole passage. Some mathematics text seems to stress obvious points while omitting important steps which the writer takes for granted, though their absence may prevent the reader from comprehending what is written. It is comparatively easy for the teacher to discover these gaps in the material he writes for his own classes, and to remedy omissions orally. The task of the published writer is more difficult, and the teacher should not take it for granted that, simply because material is published, it contains all the necessary explanation; this is especially important when pupils are working individually from written materials. Clarity and flow of meaning are perhaps the most important features which enable pupils to 'get the meaning from the page', and so to read mathematics with understanding.

It is not easy for a writer consciously to bear in mind all the guidelines suggested in this chapter, and to do so might produce a stilted style of writing. However, checklists may usefully be employed when the writer is revising his work, and the suggestions contained in them may gradually become part of his natural style.

11 Teachers' use of written materials

The inquiry

Published texts are easily available for analysis, but suggestions on how teachers might improve their use of written materials can only be hypothetical unless the ways in which teachers actually put them to use are known. With the aim of finding out more about how teachers incorporate textual materials into their mathematics lessons, the writing group instituted a small-scale inquiry. A group of teachers agreed to keep diaries of their mathematics lessons for a week, and to record the strategies they used and the difficulties they found in using text in the lessons. An analysis of the replies was made, and is reported here.

Structured diary forms (Figure 11.1, pages 138 and 139) were circulated, with the following instructions for their use.

> As soon as possible after a mathematics lesson on each day of a particular week, you are asked to fill in the diary form to record the way in which you used written materials, and the pupils' reactions to them, as follows:
>
> (i) details of the written materials and the pupils;
> (ii) what, if anything, you did with the class, or a group of pupils, or an individual pupil, to provide preliminary experience before the written materials were used;
> (iii) the precise instructions you gave about the use of the written materials;
> (iv) any information you can give about reading difficulties which the pupils encountered in the written materials, and the steps which you took to overcome these difficulties;
> (v) any further comments which you would like to make about the reasons why you used those particular written materials in the way you did during this lesson.

Sixty teachers responded to the inquiry by returning diary forms; the lessons described covered the age range 4–16 years. The primary teachers who replied used a variety of types of organisation (individual, group and class teaching) and a variety of texts. However, almost all the responses from secondary schools came from teachers who organised their lessons exclusively on a class teaching basis; all their textual materials were taken from the SMP series.

Much of the analysis of the diary forms, which follows, was carried out by Sister Timothy Pinner, OSU, and we are very grateful for her help in this project.

Analysis of the diary forms

The teachers who responded had very different ideas about the range of skills which should be included in 'reading'. Individual teachers seemed to focus on only one or two of the following reading skills, but each of these skills was listed by one or more teachers:

Sounding (saying) words, sentences and symbols.
Comprehension of the OE sense of the passage.
Comprehension of the ME sense of the passage.
One or more aspects of the full range of communication skills (including the child's own writing).

Responses from teachers

The infant teachers who completed diaries concentrated on oral discussion of practical work, written work not being appropriate to their style of teaching. Instructions they gave for the children to carry out were either oral or pictorial, so that picture reading skills may have been involved; however, the teachers did not comment on these skills. The children's writing usually consisted of single words or numbers; there was little stress on compiling either English or mathematical sentences. The infant teachers saw their most important language task in teaching mathematics to be the development of mathematical vocabulary.

From the beginning of the junior school years onwards, children are reading books in all subject areas, and teachers often expect them to read mathematics books. This is often a stage in which children seem 'to have no difficulty with the maths', but they struggle in interpreting instructions from written material. The teachers ascribed these difficulties to a number of causes, as the following comments show.

Understanding

'They could read the card but the words did not convey any meaning.'

'Although the children knew the words they seemed unable to comprehend the questions.'

Concentration

'There are too many questions on the one page for them to concentrate adequately.'

'Many of the reading difficulties originated from a lack of concentration.'

Vocabulary

'Specific problems with decoding the words "estimate and record" . . . when the meaning had been discussed one child commented, "Why didn't they use guess and write down?" The wording "Copy and complete" means very little to any of the children.'

(a) Page 1

DIARY FORM – please fill in one of these each day.

Date:

MATERIALS Please fill in the appropriate line below.

a. Book — Title: .... Page(s): ....

b. Published Workcard or Worksheet — Series title: .... Number: ....

c. School-made material (Please attach a copy if possible) — Topic: ....

PUPILS Please fill in the appropriate line for the event recorded below.

	Number of pupils	Age	Ability range
Class			
or Group			
or Individual	1		

Introduction and/or preliminary experience:

(b) Page 2

Instructions to pupils about the use of the written materials:

Reading difficulties and how they were overcome:

Further comments on these materials and their use:

Figure 11.1 Structured diary forms: (a), (b) and (c) show only the text from the three A4 sheets provided.

Phraseology

'She read the card easily and then said, "It's got here fill in the missing number. But if it's missing I can't, can I? Where is it?" After explanation of what to do she got on easily.'

Confidence

'There were questions from the children: "What do we have to do?" Most of them needed an explanation and reassurance; the confident ones didn't ask.'

The idea that some of the pupils' problems might be ascribed to reading difficulties seemed to be new to many of the secondary teachers. Their responses showed less awareness of language difficulties than did the responses of the general teachers of younger children, all of whom taught language as well as mathematics. A typical response from a secondary teacher concerned the use of text by pupils in a CSE group: 'There were no reading difficulties, only in the interpretation of what the question meant.'

(c) Page 3

GENERAL COMMENTS

School structure for teaching mathematics:

(e.g. single-age class, vertical grouping, setting, mixed ability, team teaching).
Please describe your situation.

Style of teaching organisation in the classroom:

(e.g. individual/groups/whole class)
Do you use one form of teaching organisation for mathematics most of the time, or do you vary your pattern? Please explain briefly.

Further comments

We should find any further comments you wish to make on the following points most helpful:

(i) the general use of written materials
(ii) the week's diary

Discussion of the results

It seems very clear from the diaries that the transition from oral to written instructions in mathematics is very significant; junior teachers can smooth this transition if they consciously employ some of the following strategies:

- Reading through the material with the children and talking about possible ways of tackling a problem.
- Providing revision of concept words through discussion, flash cards or pictorial clues.
- Helping children to develop a personal mathematical dictionary for future reference.

Another strategy which teachers often use is:

- Using written material to consolidate experience already gained.

However, if this last strategy alone is used, the children already know the instructions from their previous experience, and therefore they may be able to do the work without needing to read the instructions. Many children discover a variety of strategems for finding out how to do the mathematics without reading the instructions in the text; this does not build up good reading habits for the future. An example of such a strategem is approaching the teacher with the question, 'What do I have to do on this page?' If the teacher answers directly, then reading problems do not have to be faced for another page.

At the junior stage it is essential that the text should be at a sufficiently simple level to encourage children to read it, rather than to avoid reading it. Sometimes the text in a passage needs to be simplified, but it is not easy for a busy teacher to find time for the explicit detailed examination of the text that would identify its difficulties. Unawareness of textual problems easily leads the teacher to the strategy of explaining at the blackboard, with the instruction, 'Do the rest the same way.'

Some secondary teachers avoid reading problems in the same way; secondary mathematics text is more complicated, and involves an ever-increasing vocabulary, greater use of symbols and diagrams, an increasing use of typographical signals, together with instructions which often lack detail, and an increasing number of follow-up questions at the end of examples. At the secondary stage, 'reading' needs a broad interpretation; it includes comprehension and the associated communication skills. It is easy for the teacher to rely largely on exposition at the blackboard, and to fail to prepare the pupils for the nuances of language found in the mathematics textbook and examination question. This provides a short-term solution to the pupils' problems, but does not enable them to build up the reading skills in mathematics on which they will need to rely in the later secondary years. Textbook exercises often contain unusual and formalised language structures which call for interpretation; pupils need help with this, as well as in learning the skills of reading mathematics text. The strategies used for improving reading in mathematics must depend on the pupils' general reading skills. Some possible approaches include:

- Discussion of new vocabulary.
- Using oral work and conscious reference to synonyms to broaden mathematical vocabulary;
- Explanation or 'translation' exercises which enable the pupils to identify the formula or type of calculation to be used in a particular example.
- Giving pupils opportunities to read symbolic expressions aloud.
- Discussion of ways of recording answers to questions of the type 'What do you notice?', so that the use of mathematical language is encouraged.
- Discussion of the format of the text, so that the pupils are led to an appreciation of the meanings of the typographical signals used.

A teacher of mathematics in either a primary or a secondary school has access to only a limited number of texts, and so has to help the pupils to make the best use of what is available. If the teacher analyses the language of the text which he intends to use, and plans appropriate preparatory work before introducing the text to the pupils, they are likely to tackle the work with greater confidence. This preparatory work which is needed before pupils can read mathematical text for themselves may include:

- Preparatory work on new vocabulary and symbolism.
- Revision of vocabulary and symbolism previously used.
- Using a simplified version of the text.

During the actual reading of the text, some of the following devices may be helpful:

- Reading aloud with meaning.
- Formation of personal glossaries or dictionaries.
- Discussion of the translation of the problem into mathematical terms.
- Discussion of means of analysing problems.
- Discussion of ways of recording responses to questions.
- Discussion of alternative ways of phrasing an explanation or question.

Some of these ideas are discussed further in the next chapter.

The work of V.D. Petrova

V.D. Petrova was a Russian elementary school teacher, whose mathematics teaching was observed by Kalmykova in 1952–53; an account of her work is reprinted (in English) in Floyd (1981). Petrova was outstandingly successful in teaching young pupils to solve word problems correctly and independently, and to build upon that ability as they grew older. She worked in a very different culture and school system from ours, and more than thirty years ago, but her methods have given rise to ideas which can be adapted successfully to mathematics teaching in western countries in the 1980s.

> The pupils of V.D. Petrova . . . showed an excellent ability to solve problems without assistance. . . . When difficulty arose, V.D. Petrova's pupils returned to the text of the problem, reread it and looked through the solution of the problem they had done. They corrected most of the errors they made themselves.

With pupils in the first grade, Petrova concentrated much of her attention on teaching them how to read problems; many of her lessons were based on the reading aloud of problems.

> From the beginning the teacher continually emphasises that each word in the problem, regardless of how small it is, has its importance. If one changes 'tiny little words' – 'in' and 'on' – the entire sense of the problem is changed. . . . The teacher emphasised that the solution itself largely depends on a correct reading of the problem: 'Valya, here, read the problem poorly and cannot explain its solution; and Katya was mistaken because she missed this important little word "than" when she read the problem at home.'

The following problem is the focus of much of the description of Petrova's way of working:

> "Ten aspen logs were put into a stove, and six fewer birch logs were put in than aspen. How many logs were put into the stove in all?"

When a pupil proposed subtracting 6 logs from 10 logs, the teacher asked her to reread the part of the problem that implied that subtraction should be carried out ('six *fewer* birch logs'), so that the proposed method could be verified against the wording of the text. Petrova also encouraged her pupils to analyse their mistakes, showing that the basic cause of the mistake was often superficial reading of the text. Much of her work was oral:

> V.D. Petrova worked very hard to develop the speech powers of her pupils. She broadened their active vocabulary, including words to signify abstract concepts (weight, quantity, etc.). She constantly required detailed, precise answers, without extra words, to her questions. By the first half of the year her first-graders could, without the teacher's helping questions, transmit the content of a problem and explain how to solve it.

The emphasis in the Soviet primary mathematics curriculum of the 1950s was on formal arithmetical problems such as this, done in the abstract. However, the ability to 'transmit the content of a problem' orally, and 'explain how to solve it' is equally valuable in solving the more concrete and practically based problems which our primary mathematics curriculum uses, as well as in the more abstract problems of the secondary mathematics curriculum. Petrova was chosen for observation because she was notably more successful than most of her colleagues in Soviet schools in teaching her pupils to solve mathematical problems independently. Her pupils remained successful problem-solvers as they grew older; it seems that the skills of using language to explain what they were doing exactly and precisely, and to analyse and solve their difficulties, remained with them as a firm foundation for later mathematical understanding.

12 Improving the reading ability of the reader

Learning to read Mathematical English

The last two chapters have dealt with ways of making written materials easier to read, and with means whereby teachers may make more effective use of textual materials. In this chapter, we consider a third route in the exploration of methods of overcoming reading difficulties in mathematics. This third route is concerned with ways of teaching pupils to read mathematics with greater comprehension.

In the primary school, pupils are learning reading skills at the same time as they are learning mathematics; however, the techniques of teaching reading are seldom applied to the teaching of *mathematical* reading. Many devices used in the teaching of reading, such as flashcards, comprehension exercises and personal dictionaries, can easily be adapted to mathematics. Indeed, because most primary teachers teach both reading and mathematics, they are well placed to apply techniques they use to teach reading to help children to read mathematics more easily.

Other than in remedial programmes in secondary schools, secondary pupils do not usually receive specific reading lessons; consequently they are not specifically encouraged to develop their reading skills in various subject areas. Thus, the teaching of skills which are fundamental to the effective reading of mathematics text can only take place in mathematics lessons.

Pupils need to use many of their OE reading skills in reading ME, but additional reading skills are required for ME. These skills can be classified at three levels:

Decoding

The simplest level of reading skill is that of decoding the words and symbols on the page. The reader needs to recognise and identify the meaning of each word or symbol. For effective ME reading, the reader must also decode graphic material such as illustrations.

Arriving at meaning

Having decoded the building-bricks of each part of the text, the reader must then build up its total meaning. He needs to comprehend the information in the text, and to relate it to his existing conceptual knowledge.

Interacting with the text

The doing of mathematics is an active process, and one that produces a special relationship between the reader and the text. It demands that the reader should be active rather than passive. He needs to respond to what is written through his own active investigations.

At each of these levels, it is possible to identify exercises and to suggest strategies which may help to develop the pupils' reading skills.

Reading the words of ME

The decoding of words and symbols starts with the 'saying' of the word or symbol. When a mathematical word has been decoded, the reader needs to comprehend it and relate it to the concept presented. Workers in the USA have suggested many activities which may help pupils to read the vocabulary of ME. These activities are designed to draw pupils' attention to the meanings of the mathematical words they will encounter in their reading.

Cribb (undated) suggests the following activities:

1. *Unscrambling words*
   In this activity the pupil is, for instance, asked to unscramble:
   UEBC UAEQRS LAETNRGLI VDEIID
2. *Matching words, symbols and clues*
   In this activity, the pupil has to match the scrambled words with the correct symbolic forms, and also with the correct descriptions; the following example is given by Cribb:

Word/letters	Symbol	Clue/description
UEBC	□	Has 3 sides
UARQES	△	Has 6 faces
VDEIID	(cube)	Opposite to multiply
TAEINRGL	$\div$	Has 4 vertices

3. *Selecting the best answer*
   An example is:
   Bisect means (a) cut into pieces
   (b) cut into two pieces
   (c) cut into two equal pieces
   (d) cut twice
4. *Matching words with phrases*
   In this activity, words have to be matched to phrases which explain them; for example:

evaluate	put in a table
estimate	find the value of
tabulate	find an approximate value of

The first two of Cribb's activities are designed to aid spelling and word recognition. Although they may be useful, linguists would probably recommend an approach based on spelling patterns. The other two activities are designed to ensure that pupils comprehend the vocabulary of ME.

Earle (1976) suggests another vocabulary exercise; this is an activity using 'dominoes' which can be divided into groups of three; each group contains one mathematical symbol, its equivalent in words, and an example of mathematical expressions which might contain the symbol (Figure 12.1). The dominoes are shuffled, and pupils are asked to group them into matching sets of three.

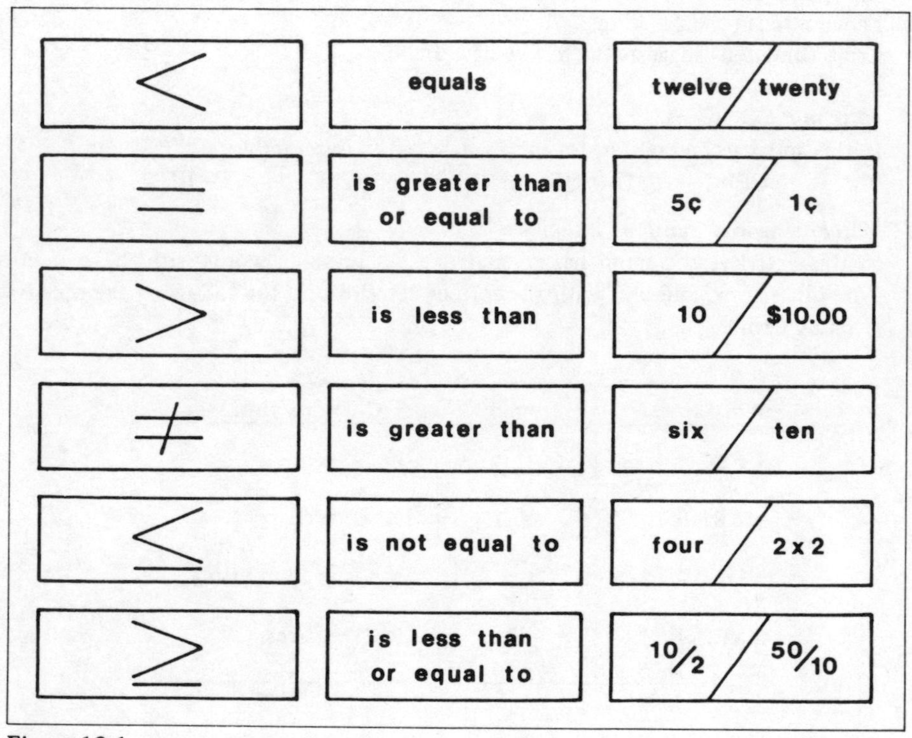

Figure 12.1

Earle also suggests the use of crosswords. The one we have constructed (Figure 12.2) is a simplification of those usually found in newspapers. Pupils should work together and make use of their dictionaries and glossaries. Earle suggests that the teacher should first list the important words in the textbook chapter which the class is about to study and that unscrambling and matching exercises, crosswords and similar activities should be devised using these words. The crossword shown in Figure 12.2 makes use of words connected with circles. This approach has the advantage of providing a thorough preparation for the work ahead, though it would be very time-consuming to carry out a programme of reading work before each chapter of a textbook. Nevertheless, such activities may be more effective than merely giving definitions of new words.

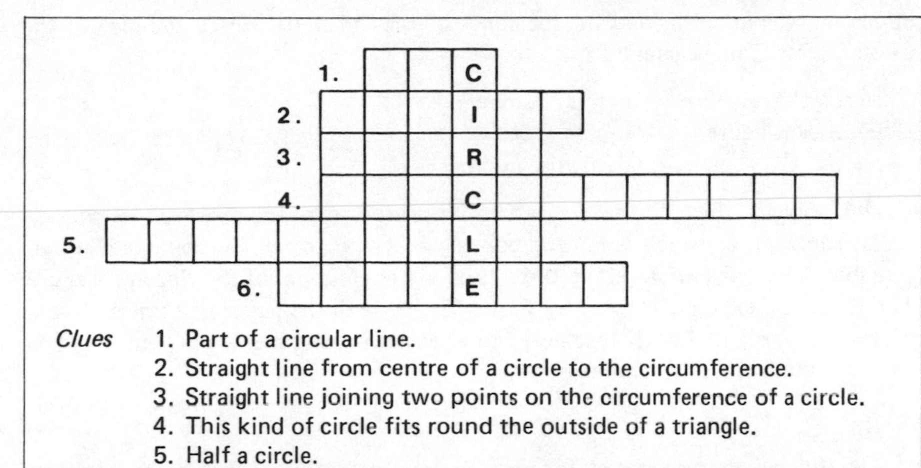

Figure 12.2

Another activity suggested by Earle draws pupils' attention to the possibility that a word or phrase may have more than one meaning. He refers to 'characteristics' rather than 'meanings' since the phrases provided are not necessarily dictionary definitions – they are aspects of the meaning of each word. The pupils are asked to tick those characteristics which apply to the word or phrase. Earle's example is shown below.

1 *a simple closed figure*

. . . . is drawn in a plane
. . . . is a polygon
. . . . encloses part of a plane
. . . . is formed by line segments
. . . . has boundaries that intersect
. . . . always has 4 or more sides

2 *a polygon*

. . . . is a simple closed figure
. . . . is not drawn in a plane
. . . . is formed by a line segment
. . . . encloses part of a plane
. . . . is like a circle

3 *lw*

. . . . is a misprint
. . . . is a formula
. . . . indicates multiplication
. . . . stands for length and width
. . . . should be printed in capital letters
. . . . can be applied to all polygons

Some authors suggest asking pupils to list the different meanings of those ME words which have more than one meaning. This is a rather routine task, although it can be useful. An interesting variation is based on the fact that two or three OE words are often put together to make a ME phrase which has a special meaning.

Pupils may be asked to find the meanings of the individual words and also of the whole phrase. Some examples are:

closed figure	natural numbers
significant figure	square root
pie chart	triangular number

One way in which the text can help its readers is by the provision of *context clues*, so exercises which stimulate pupils' ability to search for context clues are valuable. Most suggestions along these lines are adaptations of the cloze procedure. The following example is given by Earle. The style of the passage is American, and may not appeal to British teachers, but the activity is a useful one and might be translated into the British context.

Directions

The following passage was taken from your text. Every fifth word has been deleted. WITHOUT LOOKING AT YOUR TEXT, see how many sensible words you can replace. Hint: sometimes the word that makes the most sense is a very simple word, rather than a 'mathematics' word. You may be surprised at how many words you can supply that make the completed passage sensible reading. Feel free to share your ideas with your working partner, and be sure to ask him for his ideas. When you have completed as much of the 'message' as you can, refer to page 242 in your text, find out how well you did, and complete the message so the author would approve of it.

Exercise

The theorem to be proved in this section is one of the best known and most useful theorems in all mathematics. It states an ________ relationship between the lengths ________ the legs and the ________ of the hypotenuse of ________ right triangle. Most of ________ are already familiar with ________ relationship, perhaps knowing it ________ such names as the '________ Triangle Principle', the Pythagorean ________' or the 'Rule of ________'. In any right triangle ________ square of the length ________ the hypotenuse is equal ________ the sum of the ________ of the lengths of ________ legs.

When the cloze procedure is used as an exercise, there is no need for words to be deleted at regular intervals; words are deleted which have enough context clues to enable them to be identified. The following example is based on a British text.

Longer distances are measured in ________ and kilometres.
1 kilometre (1 km) = 1000 metres

The ________ from Tipton to Stanton is
2 km and 700 ________ or 2700 m.

Now ________ these:

1900 m = ☐ km and ☐ m.

Pictures could also helpfully be used in these context search activities. Pupils benefit more from cloze exercises done in groups than when they work alone; the discussion produced in a group by a cloze exercise encourages pupils to look for context clues in their reading. For example, the first space in the above cloze exercise can only correctly be filled in by a reader who uses the next line to help him; discussion will encourage pupils to point this out to one another.

Because the words used in mathematics often refer to rather sophisticated concepts, it is sometimes difficult to distinguish between teaching mathematics and teaching reading. A child who did not recognise the word 'confectioner' in a piece of OE writing could be given a simple alternative phrase 'a man who runs a sweet shop'. The child's difficulty is not in understanding the concept – he will know what 'a man who runs a sweet shop' is. In mathematics it is often difficult to explain a word merely by giving an alternative phrase; it is necessary to teach the mathematics involved. As an example of this, we look at the case of an 8-year-old child who made an error in answering the following question in his mathematics book:

> Give the largest four-digit number, using any digit only once.

His answer was '10, 98'. He had initially not recognised the word 'digit', but had found a glossary explanation:

> One of the numbers from which the larger number is made; e.g. 817 has digits 8, 1 and 7.

He then knew enough about digits to know what a 4-digit number was, but not quite enough to realise that 10 is not a digit. This example shows how conceptual understanding needs to inform the meaning of a word.

Reading symbols

The ability to read symbols is especially necessary in the reading of mathematics. If pupils are to decode symbols easily, skills such as the following are needed:

- Decoding symbols which directly replace a word or phrase; these can easily be verbalised:

$$9 + 3 = 12 \qquad 9 > 8$$

- Decoding symbols which have no clear verbal equivalent; for example, the brackets in:

$$(2 + 3) \times 4 = 20$$

- Decoding symbols which are built from spatial arrangements of simpler symbols:

$$3^2,\ 32,\ 2^3,\ \frac{4}{5},\ \frac{x^2}{2},\ 3x$$

- Decoding symbols which require a non-linear direction of reading:

$$[\tfrac{3}{4} + 5]^{2}$$

The last example may perhaps be read by first focusing on the brackets, then on the '2', then the '+', the '3', the '–', the '4', and finally the '5'. The strategies of individual readers vary, but a reader who decodes this formula strictly from left to right is unlikely to pick up the meaning very efficiently.

Kane, Byrne and Hater (1974) suggest asking pupils to draw arrows on a worksheet to indicate the directions in which they have read the words and symbols. This is not expected to be a reliable record of what pupils actually did, but discussion of the results with the teacher could help pupils to develop better reading strategies. Other suggestions for developing skills of symbol reading involve using cards with symbols on one side and words on the other, and asking children to translate symbols to words and vice versa.

Arriving at the global meaning of a passage

In the previous sections we discussed ways of obtaining meaning from individual words and symbols; however, another necessary stage is that of arriving at the meaning on a more global level. To give an idea of the importance of a global grasp of the message, we consider a pupil's error in answering a question rather similar to that discussed on page 149.

Textbook question:
Write the largest four-digit number that has the digits 1 and 9 (the same digit must not be used more than once).
Pupil's answer:
9876

Although the pupil read and fully understood all the words including the reference to 1 and 9 he just could not 'see' the significance of the 1 and 9, even when they were drawn to his attention. The message that 1 and 9 must be used was somehow lost for this pupil in the structure of the question.

In mathematics, too, the *form* of the presentation is often used to communicate a message, so that reading the individual symbols is not enough. As an example, we consider the following two number sentences:

$$3 + 15 = \square$$

$$24 \div 3 = \square \quad \times 4$$

It is easy to see why a child could fill in the answer '8' to complete the second sentence. A box is often used in text for young children to indicate, as in the first example, that a calculation is to be done and the answer written down. Interpreting the second sentence in the same way, and reading from left to right, the pupil completes what appears to him to be the first calculation, and then perhaps ignores the calculation of 8 × 4, for which a box is not provided. Thus, he needs to comprehend the sentence in this example as a whole, so that the missing number can be filled in to make the sentence true.

Our next example is taken from a text for 8-year-olds:

> Take the sum of the even numbers between 1 and 5 from the sum of the odd numbers between 2 and 6.

Even though the individual words are easily understood, a group of children whom one of us observed found the complete sentence incomprehensible. The source of the difficulty is largely in the syntax of the sentence. A teacher can break the sentence down into individual questions such as:

> What are the even numbers between 1 and 5?
> What is their sum?

But the pupil who attempts the question by himself faces a difficult problem in comprehending what he is to do. One teaching technique is to ask pupils to underline action words such as *take* and *from*. This provides the basis for questions: 'Take *what*?' '*From* what?' Although such questions are linked to the particular example, they may suggest to the pupil ways of working which will help him when he reads other problems.

Earle gives examples of activities designed to help at this level of reading. One exercise asks the pupil to decide whether a particular statement is actually made in the text.

> Which of these statements are LITERALLY STATED by the author?
> – On the number line, all points to the left of zero are positive.
> – 5 is an example of a number that is to the left of zero on the number line.
> – On the number line, any number is less than every number to its right.

Earle's next exercise helps the pupil to question himself about what he is reading. Again, the activity relates to a particular problem in the textbook and pupils are asked to say which information is given in the text.

> – The Browns had cake for dessert.
> – They ate the whole dessert.
> – They ate $\frac{2}{3}$ of the dessert.
> – They had only $\frac{3}{4}$ of a dessert to start.
> – They ate three-fourths of it.
> – They started with two-thirds of a pie.
> – Dinner lasted three-quarters of an hour.

The following exercise given by Earle is intended to develop reading for comprehension.

Mary bought 2 yards of ribbon at 20 cents a yard and some cloth for 50 cents. How much change did she get from a dollar bill?

A. True or false? If an answer is false, cross out the wrong word or words and correct them.

.... Mary bought 2 yards of ribbon.
.... The ribbon costs 15 cents a yard.
.... The cost of 2 yards of ribbon was 50 cents.
.... She bought 2 yards of cloth at 50 cents a yard.
.... The price of the cloth was 50 cents.
.... She paid the clerk with a five dollar bill.

B. Underline the question in the problem.

C. Check (√) the correct answer below.
In solving this problem you are asked to find:
.... How much change she received.
.... The price of ribbon and cloth.

Research on word problems

Much of the research carried out in the USA at this level of reading has been concerned with so-called *word problems*; these are mathematical problems presented almost entirely in ordinary English words. Many are arithmetical problems which are rather similar to many of those used in primary and middle schools in Britain.

The first task a pupil has to carry out in reading a word problem is to absorb the basic information it conveys. Petrova's approach (see Chapter 11, p. 142) is to ask the pupils to read the problem out loud and to convey the meaning through the manner of reading. This helps in attaining meaning and can be internalised when the pupil is reading silently.

Davidson (1977) reports on the success of an approach in which children were invited to make up story problems for themselves. He and his students noticed that many pupils produced rather strange problems. One of the students, for example:

> .... was amazed when every child in the room answered 7 to this problem:
> Jim has 3 sisters, Bob has 3 sisters, Joe has 4 sisters. How many girls in all?
> Again, the children knew the situation. Jim and Bob were brothers.

The children did not always include all the relevant information in the problems they made up, particularly if the information was obvious to them. However, this could be remedied, and the children shown what was missing. The use of this technique could help children to realise the ways in which information is conveyed in the words of the problem.

Asking pupils to invent problems is one way of helping them to become familiar with the general style of word problems. However, there are other ways of helping pupils to read more analytically, by offering guidelines and strategies. Earp (1970) outlines a general reading strategy:

1. Use a first reading to visualise the situation . . .
2. Reread to get the facts, paying particular attention to the information given and the *key* question . . .
3. Note problem vocabulary or concepts and explore these with the help of the teacher.
4. Reread as help in planning the steps for solving the problem.
   On this reading some arithmetic authorities have the child state the situation in a mathematical sentence. A sensible estimate of the answer may be made . . .
5. Read the problem again to check your procedure and solution.

Other writers describe in more detail some techniques for developing the kind of strategy suggested above. Henney (1971) describes a method in which the pupil has a worksheet divided into the sections given below; he has to fill in a chart, basing the answers on a particular word problem.

Verbal problem	Relation sentence
Main idea	Mathematical sentence
Question	Computation
Important facts	Answer sentence

We do not wish to encourage a rigid formula for reading word problems, but work based on these ideas might help pupils to see what to look out for when they are reading a problem. Riley and Pachtman (1978), in an article entitled 'Reading Mathematical Word Problems: Telling them what to do is not telling them how to do it', describe the use of reading guides. They define a reading guide as a set of questions or activities which pupils work on; this work helps pupils to comprehend the passage. For instance, portions of the word problem are presented, and the pupil has to tick those which are helpful; mathematical statements are presented, and the pupil has to select those which seem relevant; various computations are listed and the pupil seeks the appropriate one. At various levels, the reader is encouraged to survey relevant data and to make decisions about what to do. Again, the technique is a form of training rather than an algorithm for reading: the guide is intended to induce the reader into a simulation of good reading habits. It is hoped that these habits will persist and help him in other situations.

The ability to choose the correct mathematical technique to solve a word problem is closely related to the ability to read the problem and to form a mental picture of the information given. Brown's report of her research on Number Operations, in Hart (1981), reveals how difficult it is for children to decide on a mathematical model for a word problem. Choosing the correct mathematical model – for example, choosing to do 16×21 when asked for the area of a lawn 16 m by 21 m – is a mathematical skill rather than a reading skill. However, techniques for helping children to read a word problem must include some reference to the process of making the mathematical model. Henney's scheme, described above, includes making a mathematical model under the heading 'mathematical sentence'. It also includes the process of translating from the mathematical model back to the original situation, under the heading 'answer sentence'.

Research and development are needed on the construction and evaluation of suitable activities for British pupils, either at the level of developing specific reading skills for ME or at the more general level of stimulating the use of active reading styles.

Reading and mathematical activity

The purpose of mathematical text is to develop some form of mathematical thinking in the reader. Hence the manner of reading in mathematics must involve the reader in mathematical activity.

Text is at a great disadvantage as a teacher of mathematics: it cannot interact with the pupil. A live teacher can present information interactively, or ask a pupil to solve a problem and react immediately to his attempt. The printed page cannot do this. This is a serious difficulty, because mathematical writing aims to go further than merely communicating information; it tries to make the reader think, to help him to develop the way he thinks, and to enable him to do some mathematics.

Brookes *et al.*, in an ATM pamphlet on *Language and Mathematics* (1980), discuss the way language is used in mathematical activity:

> There is an aspect of doing mathematics which is more like writing a poem than it is like talking. When trying to write a poem one struggles with the possibilities and consequences of particular phrases; when trying to work at a piece of mathematics, one struggles with the possibilities and consequences of choosing certain signs, or a sign-system; in both cases one is trying to capture awareness.

Formal polished mathematical writing occurs at a later stage in the development of a piece of mathematics than does the activity itself. To quote Brookes *et al.* again:

> There is a form of mathematical presentation which is a by-product of the creative or problem-solving process. When tidying up their work, mathematicians may invent some clearly defined terms and axioms, and proceed to demonstrate what has to follow, given such starting points. Such formalised demonstrations can only be really useful to someone who has access to the mathematical actions to which they refer, though at a sophisticated level the very form may assist access.
>
> Unfortunately, textbooks and teachers frequently use tight and formal language when they present mathematics to children. Such a presentation appears to be an echo of the formal demonstration. It gives no hint of the language of struggle through which mathematics is created and owned, whether by adults or children.
>
> . . . However sensitively composed, the language and symbols of a textbook or work-card can only be an invitation to mathematical action; clearly they cannot guarantee that it will happen.

In the classroom, the presence of the teacher should ensure that pupils attempt mathematical activities following their reading; however, they need to be helped to realise that the actual reading of mathematics is a questioning, active, interactive process. For instance, when confronted by a worked example, pupils should not expect merely to read through it. They may cover up the solution and attempt the problem themselves. Then when they get stuck, they can peep at the solution for clues. Finally, they can study the worked example in the light of their own experience, compare its thinking with their own and ask themselves about the reasons for any differences.

This questioning, active approach to mathematical reading is often encouraged in higher education, when ME texts are often very formal in language and presentation. Then it is certainly not enough to read passively; the reader needs to be mathematically involved. Green and Webster, in *Managing Mathematically* (1976), address a special preface to the reader, who is probably a student in higher education:

> To read this book you need:
> a pen
> a lecturer
> a library
> Mathematics is an active pursuit, like swimming. You would not expect to be able to swim after reading a book; you need to try out the instructions and through these to learn new skills.

Brunner (1976) analyses the abilities required of students in reading a mathematics text at the higher education level. In describing the skills needed to learn a definition, she writes:

> When a reader is confronted with new definitions, he is apt to engage in a conversation with himself. He reasons about the definition. He works with pencil in hand and provides his own additional examples, illustrations and extensions. Mathematical exposition does not reward the reader who does not work through the reading.

In her research, Brunner identifies some of the skills associated with this involvement with the text. Although these are not relevant in their entirety to pupils at school, the identification of parallel skills at school level will be an important ingredient in the search for ways to encourage pupils to interact more with their mathematics textbooks.

Kane, Byrne and Hater (1974) list some skills associated with mathematical reading, one of which they call 'Reading with Paper and Pencil'. They distinguish this from note-taking; they are in fact advocating an interactive approach to text. They suggest that teachers might try an experiment in which half the class use pencil and paper while reading and the other half do not; the teacher could then observe which half had made better progress. Having pencil and paper is, however, only half the battle: pupils also need strategies for using them effectively. These include:

- asking themselves questions about the text,
- trying to answer these questions in writing,
- drawing diagrams,
- investigating mathematical results related to those in the text.

The style of write-in worksheets, and that used in some books, is intended to encourage interactive reading. However, the attempt to promote interaction may distort the syntax and layout of the text, and may introduce 'rhetorical' questions whose desired response is unclear. It is of more lasting value to the pupil to learn how to read mathematics. This is no easy task, and can be achieved only by long-term persistence in developing active reading skills, with the encouragement of a teacher who realises the importance of reading for learning in mathematics.

13 Directions for research

Introduction

It is not our intention in this chapter to suggest a list of possible research projects. Rather, some general research directions will be indicated that suggest a framework for investigation. We are concerned also that the professional aspects of the teacher's work on the reading of mathematics should be developed, and that teachers and writers should be able to contribute to the lines of enquiry, and to benefit from the results. Thus, the word 'research' is to be understood in its widest sense in this chapter.

Much useful research may be carried out by a teacher in the classroom. Indeed, comparison of the readability of texts which are being considered for use, the investigation of children's reading problems, and the evaluation of techniques for teaching the reading of mathematical text, are all a part of a teacher's daily work, and they constitute informal small-scale research activity.

More formal research leading to published results can run parallel to this informal work. Formal and informal research activities are closely linked; the spread of informal research is assisted by the availability of published reports from formal research projects. The same people may also be involved in both formal and informal research.

As can be seen from the bibliography at the end of this book, a substantial body of work already exists; much of it was carried out in the USA. In Britain, there has not been the same emphasis on the readability of texts, and on reading in the subject areas, as there has been in the USA. Much British research has concerned 'language' in a more general way; for example, there is considerable work on the role of language in concept formation, the teacher's use of spoken language, and the role of everyday language in mathematics. Much of this work is discussed in Austin and Howson's (1979) review of the literature in the field of language and mathematics; readability forms a small part of this survey. Hence, research into reading in mathematics, and into the readability of mathematical texts, particularly those used in British schools, is rich in potential avenues of exploration.

We now consider the following three broad areas:

1. the development of readability formulae;
2. the establishment of criteria for the evaluation of written material;
3. the evaluation of methods of helping children to improve their reading of Mathematical English.

Readability formulae

Readability formulae have been the subject of research over many years, as described in Chapter 7. The problems involved in creating a reliable readability formula are clearly identified in the literature, and further work needs to be done in the light of this knowledge. In particular, since so much of the work has been with American children and American texts, there is a need to investigate similar formulae based on the study of British books and pupils. The development of British versions of the Kane, Byrne and Hater (1974) formulae would be a useful piece of work, for example. Such a development could involve three aspects:

1. Cloze tests have been used to calibrate the readability of passages of ME. This method of deciding on their reading difficulty needs further investigation.
2. British lists of familiar and unfamiliar mathematical and non-mathematical words and symbols would be valuable. These should be constructed by investigation with pupils in order to identify the words and symbols they actually know, as well as by inspecting commonly used texts.
3. The variables which correlate with the readability of standard passages may be different for British and American text. Not only are the lists of familiar words likely to be different, but the styles of writing in British and American texts are quite different.

It should be remembered that although readability formulae are potentially useful for comparing texts, they are severely limited in their usefulness for such tasks as establishing the reading age of a text, identifying particular reading problems, and helping authors to make their writing more readable. This is because factors such as sentence length and number of question marks, which are associated with reading difficulty, do not *cause* the difficulty; they merely correlate statistically with the difficulty. However, such correlations can pose questions which may lead to a better focused investigation of the causes of reading difficulty.

Criteria for the evaluation of written material

A significant theme in earlier chapters is that the informed judgement of a teacher or writer is likely to be the most reliable way of evaluating written materials in mathematics. However, we have suggested that such judgements may be better informed if criteria such as those set out in Chapters 2 to 8 are used. These criteria indicate probable causes of reading difficulties, and they are therefore likely to be helpful for practical purposes. Hence, a closer evaluation of these criteria is a desirable direction for enquiry. Some examples of possible investigations follow:

1. Because unfamiliar words are liable to cause reading difficulties, further investigation to find which words and symbols are familiar to British pupils should be very useful.
2. Since syntax is thought to be an important factor in readability, a study of the grammatical structure found in mathematical texts would be valuable, as would observation of children reading text containing different structures.

3. Many features peculiar to Mathematical English, such as the reading of symbols, graphic language, and the layout of the page, are not fully understood. Comparative evaluation of texts which vary in the ways in which they incorporate such features would be a way forward in identifying more clearly how these features affect readability.

As well as considering particular causes of poor readability, it is important to assess the professional value of such knowledge for teachers. The approach used in this book suggests a strategy for judging readability. First, the teacher should clearly identify the purpose of each of the different types of writing on the page, and then consider the small-scale and large-scale features contributing to readability. 'Small-scale' features are vocabulary, syntax, symbols and details of diagrams. 'Large-scale' features are those which affect the comprehension of the page or pages as a whole: the flow of meaning, layout of the page, and the interrelation of graphic material and text.

How useful would such a strategy be in carrying out the day-to-day work of a teacher or writer? Would the adoption of this strategy mean that a group of people would agree more closely about the relative readability of two pieces of text? Would this strategy enable a page to be rewritten so that pupils could read it more easily? Would a structured awareness of the features of the text help teachers to identify effective techniques for use in mathematics lessons? There are many questions such as these which could usefully be investigated in order to validate some of the theoretical study of mathematics text which we have carried out, and which we discussed in Chapters 2 to 8.

Teaching the reading of Mathematical English

One of the most important areas for development is that of helping pupils to develop skills in reading Mathematical English. A number of possible methods of doing this have been suggested in Chapter 12; these methods operate at different levels. Their effectiveness needs to be investigated, and other methods may need to be devised. Inevitably, the evaluation of such methods must be carried out over a fairly long period of time. However, informal trials of such approaches suggest that marked improvements in pupils' ability to read mathematical text may be possible. Structured work in this area requires a carefully designed approach and considerable experimentation.

Texts quoted as examples

Backhouse, J.K., Houldsworth, S.P.T., Cooper, B.E.D., *Pure Mathematics*, Longman, 1963

Burke, P., Bronto Books, Longman, 1979

Channon, J.B., McLeish Smith, A., Head, H.C., *New General Mathematics Revision*, Longman, 1976

Department for Education and Science, National Primary Survey, 1978

Fletcher, H., Howell, A., Walker, R., *Mathematics for Schools* (first edition), Addison-Wesley, 1971

Fletcher, H., Howell, A., Walker, R., *Mathematics for Schools* (second edition), Addison-Wesley, 1980

Goddard, T.R., Grattidge, A.W., *Beta Mathematics*, Schofield and Sims, 1969

Holt, M., *Maths 3*, Macmillan, London and Basingstoke, 1974

Holt, M., Rothery, A., *Mathsworks, Book One*, Longman, 1979

Holt, M., Rothery, A., *Mathsworks, Book Two*, Longman, 1979

Holt, M., Rothery, A., *Mathsworks, Book A*, Longman, 1982

Illich, I., *After Deschooling, What?*, Writers and Readers Publishing Co-operative, 1974

Ormell, C., Schools Council, *Mathematics Applicable* Project, 1978

Porter, R.I., *Further Elementary Analysis*, Bell, 1951, 1970

School Mathematics Project, *SMP Book B*, Cambridge University Press, 1969

School Mathematics Project, *SMP Book C*, Cambridge University Press, 1969

School Mathematics Project, *SMP Book D*, Cambridge University Press, 1970

School Mathematics Project, *SMP New Book 3*, Cambridge University Press, 1981

School Mathematics Project, *Revised Advanced Mathematics*, 1973

Schools Council Sixth Form Mathematics Project, *Mathematics Applicable: Polynomial Models*, Heinemann Educational Books, 1978

Scottish Mathematics Group, *Modern Mathematics for Schools*, Blackie Chambers, 1971

Scottish Primary Mathematics Group, *Primary Mathematics*, Heinemann Educational Books, 1975

Stanfield, J., *Maths Adventure 3*, Evans and Bell & Hyman, 1971

Stanfield, J., *Maths Adventure 4*, Evans and Bell & Hyman, 1977

Stewart, I., Tall, D., *The Foundation of Mathematics*, Oxford, 1977

Sylvester, J.E.K., *Mainstream Mathematics*, Book 2, Thomas Nelson & Sons, 1979

Bibliography

Aiken, L.R., 'Verbal Factors in Mathematics Learning: A Review of Research', *Journal for Research in Mathematics Education,* **2**, 1971

Aiken, L.R., 'Language Factors in Learning Mathematics', *Review of Educational Research,* **42**, 3, 1972

Assessment of Performance Unit, *Mathematical Development, Secondary Survey Report No. 2,* HMSO, 1981

Association of Teachers of Mathematics, *Language and Mathematics,* ATM, 1980

Austin, J.L., Howson, A.G., 'Language and Mathematical Education', *Educational Studies in Mathematics,* **10**, 1979, pp. 161–97

Balow, I.H., 'Reading and Computation Ability as Determinants of Problem Solving', *Arithmetic Teacher,* **11**, 1964, pp. 18–22

Bental, L.H., 'Language Difficulties in Teaching Mathematics', *Forward Trends,* **16**, 1972

Bishop, A.J., 'Actions Speak Louder than Words', *Mathematics Teaching,* 1974

Blankenship, C., Lovitt, T.C., 'Story Problems: Merely Confusing or Downright Befuddling?', *Journal for Research in Mathematics Education,* 1976, pp. 290–8

Bormuth, J.R., 'Readability – A New Approach', *Reading Research Quarterly,* **1**, 1966

Bormuth, J.R., 'Comparable Cloze and Multiple Choice Comprehension Test Scores', *Journal of Reading,* **10**, 1971

Bormuth, J.R., Manning, J., Carr, J., Pearson, D., 'Children's Comprehension of Between- and Within-Sentence Syntactic Structures', *Journal of Educational Psychology,* **61**, 1970

Bortnick, R., 'An Instructional Application of the Cloze Procedure', *Journal of Reading,* **16**, 1973

Botel, M., 'The Study Skills in Mathematics', in Melwick and Merritt, *The Reading Curriculum,* University of London Press, 1972

Botel, M., Dawkins, J., Granowsky, A., 'A Syntactic Complexity Formula', in MacGintie, W.H. (ed.), *Assessment Problems in Reading,* International Reading Association, 1973

Brown, M., in Hart (1981)

Brunner, R.B., 'Reading Mathematical Exposition', *Educational Research,* **18**, 3, 1966, pp. 208–13

Brunner, R.B., 'The Construction and Construct Validation of a Reading Comprehension Test of Mathematical Exposition', *Journal of Structured Learning,* **4**, 4, 1977

Brush, L.R., Brett, L.S., Sprotzer, E.R., 'Children's Difficulties on Quantitative Tasks; Are They Simply a Misunderstanding of Relational Terms?', *Journal for Research in Mathematics Education,* **9**, 2, 1978, pp. 149–51

Bullock, A., *A Language for Life,* HMSO, 1975

Burt, C., *The Psychology of Printing,* OUP, 1946

Call, R.J., Wiggin, N.A., 'Reading and Mathematics', *The Mathematics Teacher,* **LIX**, 2, 1966

Campbell, P.F., 'What do Children See in Textbook Pictures?', *Arithmetic Teacher,* **28**, 5, 1981, pp. 12–16

Cashdan, A., *Language, Reading and Learning,* Blackwell, 1979

Cazden, C.D., *Child Language and Education,* Holt, Rinehart and Winston, 1972, pp. 23–28

Chapman, L.J. (ed.), *The Reader and the Text,* Heinemann, 1981

Clark, F.R., 'The Problem of Reading Instruction in Mathematics', in Mazurkiewicz, A.J. (ed.), *New Perspectives in Reading Instruction,* Pitman, 1964

Cleland, D.L., Toussaint, I.H., 'Interrelationships of Reading, Listening, Arithmetic, Computation and Intelligence', *Reading Teacher,* **15**, 1962, pp. 228–31

Cockcroft, W.H., *Mathematics Counts,* HMSO, 1982

Collier, C.C., Redmond, L.A., 'Are You Teaching Kids to Read Mathematics?', *Reading Teacher,* **27**, 8, 1974, pp. 804–8

Cooper, F.F., 'Maths as a Second Language', *The Instructor,* **81**, 1971

Cormack, A., *A Study of Language in Mathematics Textbooks,* M.Phil. Thesis, London, 1978

Crabbs, L.M., McCall, W.A., *Standard Test Lessons in Reading,* New York Bureau of Publications, Teachers' College, Columbia University, 1926

Cribb, P., 'Learning Mathematical Language', *Vinculum,* The Maths. Association of Victoria, **13**, 4, undated

Dahmus, M.E., 'How to Teach Verbal Problems', *School Science and Maths,* 1970, pp. 121–38

Dale, E., Chall, J.S., 'Formula for Predicting Readability', *Education Research Bulletin,* **27**, 1948

Davidson, J.E., 'The Language Experience Approach to Story Problems', *Arithmetic Teacher,* **25**, 1, 1977

D.E.S., *National Primary Survey,* HMSO, 1978

D.E.S., *Mathematics 5–11,* HMSO, 1979

Dolgin, A.B., 'Improvement of Mathematical Learning through Reading Instruction', *High School Journal,* **61**, 2, 1977

Dresher, R., 'Training in Mathematics Vocabulary', *Educational Research Bulletin,* 1934, pp. 201–4

Dunning, R., *The Technique of Clear Writing,* McGraw Hill, 1952

Earle, R.A., *Teaching Reading and Mathematics,* International Reading Association (USA), 1976

Earp, N.W., 'Observations on Teaching Reading in Mathematics', *Journal of Reading,* **13**, 1970

Earp, N.W., 'Procedures for Teaching Reading in Mathematics', *Arithmetic Teacher,* **17**, 1970, pp. 575–9.

Earp, N.W., 'Problems of Reading in Mathematics', *School Science and Maths,* 1971

Earp, N.W., Tanner, G.W., 'Mathematics and Language', *Arithmetic Teacher,* **28**, 1980

Edwards, R.P.A., Gibbon, V., *Words Your Children Use,* Burke Books, 1964, 1973

Farr, J.N., Jenkins, J.J., Paterson, D.G., 'Simplification of the Flesch Reading Ease Formula', *Journal of Applied Psychology,* **35**, 1951

Fay, L., 'Reading Study Skills: Maths and Science', in Melwick and Merritt, *The Reading Curriculum,* ULP, 1972

Freeman, G.F., 'Reading and Mathematics', *Arithmetic Teacher,* **20**, 7, pp. 523–9.

Feliciana, S.D., Powers, R.D., Kearl, B.E., 'The Presentation of Statistical Information', *A. V. Communication Review,* **11**, 1963

Flesch, R.F., 'A New Readability Yardstick', *Journal of Applied Psychology,* **32**, 1948

Floyd, A. (ed.), *Developing Mathematical Thinking,* Addison-Wesley, 1981

Foster, J., *School books: their language, presentation and use,* Council of Subject Teacher Associations Paper 2, 1978.

Fry, E.B., 'The Readability Graph Validated at Primary Level', *Reading Teacher,* **22**, 1969

Fry, E.B., 'Fry's Readability Graph; Clarifications, Validity', *Journal of Reading,* **21**, 1977

Gilliland, J., *Readability,* Unibooks, 1972

Gilliland, J., 'The Use of Cloze Procedure in the Measurement of the Readability of Schools Council Humanities Project Materials', *Reading,* **6**, 1972

Gilliland, J. (ed.), *Reading: Research and Classroom Practice,* UKRA/Ward Lock, 1977

Grant, M., 'The Development of the Concept of Notation', *British Journal of Educational Psychology,* **40**, 1, 1970, pp. 81–2

Grouws, D.A., Robinson, R.D., 'Some Ideas Concerning the Readability of Classroom Mathematical Materials', *School Science and Mathematics,* 1973, pp. 711–16

Hafner, L., 'Research for the Classroom: Cloze Procedure', *Journal of Reading,* **9**, 1966

Hamrick, K.B., 'Oral Language and Readiness for the Written Symbolisation of Addition and Subtraction', *Journal for Research in Mathematics Education,* **10**, 3, 1979, pp. 189–94

Hanley, A., 'Verbal Mathematics', *Maths in School,* **7**, 4, 1978

Harkin, J.B., Rising, G.R., 'Some Psychological and Pedagogical Aspects of Mathematical Symbolism', *Educational Studies in Mathematics,* **5**, 1974

Harrison, C., 'Assessing the Readability of School Texts', in Gilliland (1977)

Harrison, C., *Readability in the Classroom,* Cambridge University Press, 1980

Hart, K.M. (ed.), *Children's Understanding of Mathematics: 11–16,* John Murray, 1981

Harvey, R., Kerslake, D., Shuard, H., Torbe, M., *Language Teaching and Learning: Volume 6. Mathematics,* Ward Lock Educational, 1982

Hater, M.A., *The cloze procedure as a measure of the reading comprehensibility and difficulty of mathematical english,* unpublished doctoral thesis, Purdue University, 1969

Hater, M.A., Kane, R.B., Byrne, M.A., 'Building Reading Skills in the Mathematics Class', *Arithmetic Teacher,* 1974, pp. 662–8

Hater, M.A., Kane, R.B., 'The Cloze Procedure as a Measure of Mathematical English', *Journal for Research in Mathematics Education,* **16**, 6, 1975, pp. 121–7

Heddens, J.W., Smith, K.J., 'The Readability of Experimental Mathematics Materials', *Arithmetic Teacher,* **11**, 6, 1964, pp. 391–4

Heddens, J.W., Smith, K.J., 'The Readability of Elementary Mathematics Books', *Arithmetic Teacher,* **11**, 7, 1964, pp. 466–8

Henney, M., 'Improving Mathematics Verbal Problem-Solving Ability through Reading Instruction', *Arithmetic Teacher,* **18**, 4, 1971, pp. 223–9

Holtan, B.D., Knifong, J.D., 'An Analysis of Children's Written Solution to Word Problems', *Journal for Research in Mathematics Education,* **7**, 1976

Irish, E.H., 'Improving Problem-Solving by Improving Verbal Generalisation', *Arithmetic Teacher,* **11**, 1964, pp. 169–175

Jerman, M., Rees, R., 'Predicting the Relative Difficulty of Verbal Arithmetic Problems', *Educational Studies in Mathematics,* **4**, 1972

Jerman, M., Mirman, C.S., 'Linguistic and Computational Variables in Problem-Solving in Elementary Mathematics', *Educational Studies in Mathematics,* **5**, 1974

Johnson, H.C., 'The Effect of Instruction in Mathematical Vocabulary upon Problem Solving in Arithmetic', *Journal of Educational Research,* **38**, 1944, pp. 97–110

Johnson, T.D., *Reading: Teaching and Learning,* Macmillan, 1973

Johnson, D.A., 'The Readability of Mathematics Books', *The Mathematics Teacher,* 1957, pp. 105–110

Johnstone, A., Cassels, J., 'What's in a word?', *New Scientist,* 1978, pp. 432–4

Jongsma, E., *The Cloze Procedure as a Teaching Technique,* International Reading Association, 1971

Jones, R.J.McC., *The Usability of Mathematics Textbooks as Found in Third-Year Junior Classes,* M.Ed. Thesis, Sheffield, 1976

Jones, S., *Design on Instruction,* Training Information Paper 1, Dept. Employment and Productivity, HMSO, 1968

Kane, R.B., 'The Readability of Mathematical English', *Journal of Research in Science Teaching,* **5**, 1967, pp. 296–8

Kane, R.B., 'Readability of Mathematics Textbooks Revisited', *The Mathematics Teacher,* **63**, 1970, pp. 579–81

Kane, R.B., Byrne, M.A., Hater, M.A., *Helping Children Read Mathematics,* New York American Book Co., 1974

Kerslake, D., in Hart, K., *Children's Understanding of Mathematics: 11–16,* John Murray, 1981

Kieran, K., 'Concepts associated with the equality symbol', *Educational Studies in Mathematics,* **12**, 3, 1981, pp. 317–26

Kilpatrick, 'Observations of V.D. Petrova's Class', in *Soviet Studies in the Psychology of Learning and Teaching Mathematics,* now in Floyd (1981)

King, *All Things Bright and Beautiful,* 1978

Klare, G.R., *The Measurement of Readability,* Iowa State University Press, 1963

Koenke, K., 'Another Practical Note on Readability Formulas', *Journal of Reading,* 1971

Kulm, G., 'Sources of Reading Difficulty in Elementary Algebra Textbooks', *Mathematics Teacher,* **66**, 1973, pp. 644–52

Lees, F., 'Mathematics and Reading', *Journal of Reading,* 1976

Levie, W.H., Dickie, K.E., 'The Analysis and Application of Media', in Travers, R.M.W. (ed.), *Handbook of Research on Teaching, Vol. 2,* Rand McNally, 1973

Linville, W.J., 'Syntax, Vocabulary and the Verbal Arithmetic Problem', *School Science and Mathematics,* 1976

Love, E., Tahta, D., 'Language Across the Curriculum: Mathematics', *Mathematics Teaching,* 1977

Lunzer, E., Gardner, K., (ed.), *The Effective Use of Reading,* Heinemann, 1979

Lyda, W.J., Duncan, F.M., 'Quantitative Vocabulary and Problem Solving', *Arithmetic Teacher,* **14**, 1967, pp. 289–91

MacGibbon, H., *School books: their language, presentation and use; a publisher's view,* Council of Subject Teaching Associations Paper 3, 1978

McLaughlin, H., 'Smog Grading – A New Readability Formula', *Journal of Reading,* 1971

McLeod, J., 'The Estimation of Readability of Books of Low Difficulty' *British Journal of Psychology,* **32**, 1962

Maginnis, G.H., 'The Readability Graph and Informal Reading Inventories', *Reading Teacher,* **22**, 1969

Malter, M., 'The Ability of Children to Read Conventialized Diagrammatic Symbols', *Journal of Educational Psychology,* **39**, 1948, pp. 17–34

Matthews, J., 'An investigation into subtraction', *Educational Studies in Mathematics,* **12**, 3, 1981, pp. 327–38

Meredith, P., *Learning, Remembering and Knowing,* EUP, 1961

Merritt, J. (ed.), *Reading and the Curriculum,* Ward Lock Educational, 1971

Morris, R.W., 'The Role of Language in Learning Mathematics', *Prospects: Quarterly Review of Education,* **8**, 1, 1978

Moyle, D., *The Teaching of Reading,* Ward Lock, 1968

Mugford, L., 'A New Way of Predicting Readability', *Reading,* **4**, 1970

Munro, J., 'Language Abilities and Maths Performance', *Australian Journal of Remedial Education,* **9**, 3, 1977

Munro, M.E., *The Language of Mathematics,* University of Michigan Press, 1963

Nesher, P., Teubal, E., 'Verbal Cues as an Interfering Factor in Verbal Problem-Solving', *Educational Studies in Mathematics,* **6**, 1974

Nesher, P., 'Three Determinants of Difficulty in Verbal Arithmetic Problems' *Educational Studies in Mathematics,* **7**, 1976

Nesher, P., Katriel, T., 'A Semantic Analysis of Addition and Subtraction Word Problems in Arithmetic', *Educational Studies in Mathematics,* **8**, 1977, pp. 2341–50

Neville, M.H., Pugh, A.K., 'Reading Ability and Ability to Use a Book', *Reading,* **9**, 3, 1975

Nicholson, A.R., 'Mathematics and Language', *Maths in School,* **6**, 5, 1977

Office of Instructional Services, Georgia Department of Education, *Reading Mathematics, a Reading Program for the 70s,* Georgia Dept. of Education, undated

Otterburn, M.K., Nicholson, A.R., 'The Language of (C.S.E.) Mathematics', *Maths in School,* **5**, 5, 1976, pp. 18–21

Pachtman, A.B., Riley, J.D., 'Teaching the Vocabulary of Mathematics Through Interaction, Exposure and Structure', *Journal of Reading,* 1978, pp. 240–4

Perera, K., 'The Assessment of Linguistic Difficulty in Reading Material', *Educational Review,* **32**, 2, 1980

Perry, D.K., 'Speed and Accuracy of Reading Arabic and Roman Numerals', *Journal of Applied Psychology,* **36**, 1952, pp. 346–7

Peters, J., 'Language and Mathematics Teaching in the Open University', *Teaching at a Distance,* 1975, pp. 31–34

Pomery, J.D., 'Pictorial Representation – Is There a Progression?', *Primary Mathematics,* **9**, 1971, pp. 75–84

Preston, M., 'The Language of Early Mathematical Experience', *Maths in School,* **7**, 4, 1978

Pribnow, J.R., 'Why Johnny Can't "Read" Word Problems', *School Science and Maths,* **LXIX**, 7, 1969

Rankin, E.F., Culhaine, J.W., 'Comparable Cloze and Multiple Choice Comprehension Test Scores', *Journal of Reading,* **13**, 1969

Rankin, E.F., 'The Cloze Procedure – Its Validity and Utility', 1959, reprinted in Farr, R., *Measurement and Evaluation in Reading,* Harcourt, Brace and Ward Inc., 1970

Reed, M., Waiman, H., 'Language Competence in Mathematics', *International Journal of Math Education in Science and Technology,* **9**, 1, 1978, pp. 31–33

Reid, J.F., *Reading Problems and Practices,* Ward Lock, 1972

Riley, J.D., Pachtman, A.B., 'Reading Mathematical Word Problems: Telling Them What to do is not Telling Them How to do it', *Journal of Reading,* **21**, 6, 1978, pp. 531–4

Robinson, H.A., *Teaching Reading and Study Strategies: The Content Areas,* Allen and Bacon, 1975

Russell, D.H., Spencer, P.L., 'Reading in Arithmetic', in *N.C.T.M. Yearbook,* NCTM, 1960, pp. 202–23

Schools Council Project, *Reading for Learning in the Secondary School,* School of Education, University of Nottingham, trial materials, 1980

Schoenherr, B., 'Writing Equations for Story Problems', *Arithmetic Teacher,* **15**, 1968

Schuyler, R.M., 'A readability formula program for use on microcomputers', *Journal of Reading,* 1982

Smith, F., 'The Readability of Sixth Grade Word Problems', *School Science and Maths,* **6**, 1971, pp. 554–562

Smith, F., 'The Readability of Junior High School Mathematics Textbooks', *Mathematics Teacher,* **LXLL**, 4, 1969

Skemp, R., *The Psychology of Learning Mathematics,* Penguin, 1971

Spooner, F., 'Hanging out the 'Cloze' Line', *Reading,* **8**, 2, 1974

Stokes, A., 'The Reliability of Readability Formulae', *Journal of Research in Reading,* **1**, 1, 1978

Strain, L.B., 'Children's Literature: An Aid in Mathematics Instruction', *Arithmetic Teacher,* **16**, 1969, pp. 451–5

Taschow, H.G., 'Reading Improvement in Mathematics', *Reading Improvement,* **6**, 1969

Taylor, W.L., 'Cloze Procedure: A New Tool for Measuring Readability', *Journalism Quarterly,* **30**, 1953

Tinker, M.A., *Bases for Effective Reading,* University of Minnesota Press, 1965

Treachy, J.P., 'The Relationship of Reading Skills to the Ability to Solve Arithmetic Problems', *Journal of Educational Research,* **38**, 1944, pp. 86–96

Twyman, M., 'A Schema for the Study of Graphic Language', in Kolers, P.A. Wrolstad, M.E., Bouma, H., *Processing of Visible Language, Vol. 1.,* New York, Plenum, 1979

Van der Linde, L.F., 'Does the Study of Quantitative Vocabulary Improve Problem Solving?, *Elementary School Journal,* **65**, 1964, pp. 143–52

Vinner, S., 'The Elimination of some Mathematical Confusions Through the Analysis of Language', *Mathematical Education for Teaching,* **3**, 2, 1970

Wagner, J., 'Learning to Read Maths', *Curriculum Review,* **16**, 5, 1977 and **17**, 1, 1978

Ward, M., 'Fish Tanks, Right Angles and Graphs – Assessment of Understanding in Mathematics', *Education 3–13,* **4**, 2, 1976, pp. 96–101

Watkins, A.E., 'The Symbols and Mathematical Structures of Mathematical English and the Reading Comprehension of College Students', *Journal for Research in Mathematics Education,* **10**, 3, 1979, pp. 216–18

Weijand, R.B., 'Readability of Mathematics Textbooks', *Journal of Reading,* **11**, 1967

Weintraub, S., 'Research: Textbooks', *Reading Teacher,* **2**, 1967

Weintraub, S., 'The Cloze Procedure', *Reading Teacher,* **21**, 1968

Westwood, P., Fitton, S., 'Assessing Readability', *Remedial Education,* **8**, 1, 1973

Whalley, P.C., Fleming, R.W., 'An Experiment with a Simple Recorder of Reading Behaviour', *Programmed Learning and Educational Techniques,* **12**, 2, 1977, pp. 120–4

Wheatley, D., 'Mathematical Concepts and Language 1937–1977', *The New Era,* **58**, 5, 1977, pp. 134–8

Widlake, P., 'The Textbooks That The Children Can't Read', *Education,* **23**, 1979, pp. 219–20

Will, C. Van Der, 'The wording of spoken instructions to children and its effect on their performance of tasks', *Educational Studies,* 1976

Williams, R., in Chapman (1981)

Wilmon, B., 'Reading in the Content Area: A New Math Terminology List for the Primary Grades', *Elementary English,* 1971

Index

References to pages on which main topics are given their principal treatment appear in italics.